Refurbishment and Repair
in Construction

Refurbishment and Repair in Construction

David Doran
BSc(Eng) DIC FCGI CEng FICE FIStructE

Consultant, Civil/Structural Engineer,
formerly Chief Structural Engineer, Wimpey plc, UK

James Douglas
BSc MRICS MBEng FHEA
School of the Built Environment, Heriot-Watt University, Edinburgh

Richard Pratley
B.Arch (Liverpool)

Architect, London

WHITTLES PUBLISHING

CRC Press
Taylor & Francis Group

Published by
Whittles Publishing,
Dunbeath,
Caithness KW6 6EY,
Scotland, UK

www.whittlespublishing.com

Distributed in North America by
CRC Press LLC,
Taylor and Francis Group,
6000 Broken Sound Parkway NW, Suite 300,
Boca Raton, FL 33487, USA

© 2009 D Doran, J Douglas, R Pratley

ISBN 978-1904445-55-5
USA ISBN 978-1-4398-0871-9

The publisher and authors have used their best efforts in preparing this book, but assume no responsibility for any injury and/or damage to persons or property from the use or implementation of any methods, instructions, ideas or materials contained within this book. All operations should be undertaken in accordance with existing legislation and recognized trade practice. Whilst the information and advice in this book is believed to be true and accurate at the time of going to press, the authors and publisher accept no legal responsibility or liability for errors or omissions that may have been made.

Every effort has been made to trace copyright holders and to obtain their permission for the use of copyright material. The publisher would be grateful if notified of any amendments that should be incorporated in future reprints or editions of this book.

Printed by Bell & Bain Ltd, Glasgow

Contents

5 Types of construction: disasters, defects and potential solutions   **80**

7 Case Studies   256

Foreword

In a recent report from the Standing Committee on Structural Safety [SCOSS], corporate amnesia was estimated at 35 years. This would imply that we have all forgotten what happened before 1974, and no memories survive of the problems that came with the use of various materials or forms of construction, which were so fresh in the minds of those of us in practice at that time. They are simply not in the consciousness of younger engineers, architects and surveyors, thus the need for this book. *Refurbishment and Repair in Construction* describes those earlier forms of construction and the related problems, provides solutions, and reviews the uses and properties of materials used in construction. Also included are generous lists of references and contacts should the reader need to conduct further research.

It would be foolish to embark on the refurbishment or repair of an existing construction without a thorough understanding of its properties, both material and structural i.e. why and how it remains standing. It is easy to become preoccupied and diverted by problems alone – what caused that crack or that distortion? why are bits falling off? can that wall be removed? will it take an extra floor? – and to omit that vital precursor or first step: to strive to fully understand what is being dealt with. Time spent trying to identify the cause of defects before the construction is fully understood is not simply wasted, but actually counterproductive. It can lead to incorrect diagnosis and thus remedial work that aggravates rather than alleviates the original problem, leading to a frustrated adviser and an unhappy client. This understanding requires the knowledge distilled from a lifetime of experience within the construction industry. Such knowledge is made available in this book and is invaluable for younger and future professionals as well as current practitioners. Much of the information contained in this book is no longer taught to students and apprentices, and covered only occasionally by Continuing Professional Development (CPD) presentations.

There is increasing recognition that those involved in maintaining, repairing and refurbishing existing constructions need to use sympathetic materials in appropriate ways (and not solely rely on suppliers' marketing literature) and do not apply modern materials which may change the environment of the existing construction. Many buildings must be able to breathe and can be irretrievably damaged by the use of modern strong impervious materials. A lack of awareness of such requirements is all too prevalent, such as repointing old porous brickwork laid in lime mortar with Portland cement mortar. The more that the sympathetic use of appropriate materials is emphasised, the more we are able to ensure durable and attractive repairs to our building stock.

It is gradually being appreciated that refurbishment of existing constructions, with minimal alterations, is a more sustainable and preferable approach than demolition and reconstruction, even if some materials from the demolition are saved and recycled. There are significant energy savings to be found through considerate repair and refurbishment compared with demolition and use of new materials. However, it is essential that the existing construction is fully understood through its material and structural properties. Professionals that lack such understanding may be reluctant to propose reuse of materials. They would prefer to work with what they know, not withstanding the inadequacies of workmanship commonly found with new materials, in contrast to an existing structure that has proven its acceptability. However, such professionals can now gain the understanding they need from this book and its extensive reference sources.

Lawrance Hurst
BSc(Eng) FCGI CEng FICE FIStructE,
Consultant to Hurst Peirce + Malcolm

Preface

Refurbishment and repair of a building or structure can be a daunting task – probably a more difficult enterprise than is generally realised. It may be necessary to reconstruct a building for modern use but to preserve its appearance to match its original style.

To achieve success it is often necessary to have a deep knowledge of the construction history and of the materials used in the original work. Seeking this knowledge can be difficult, even tedious. There is great deal of available data but a major problem is to know where and how to look for this information.

Any new work has to be carried out within a complex maze of legal frameworks which probably did not exist at the time of the original construction.

This book seeks to assist those embarking on such work, highlights possible pitfalls and suggests strategies which will minimise the risk involved.

I commend this book to practitioners and would take this opportunity to thank my co-authors James Douglas and Richard Pratley for their patience and professionalism in making their contributions to this book.

David Doran

Editor's note

(1) In this manual the male sex is used throughout. The term is intended to indicate both male and female sexes and should be construed as such.
(2) Cover illustrations by courtesy of: Abbey Pynford; Michael Wade of Dorman Long Technology; Steve Evans of Peel & Fowler; GB Geotechnics Ltd., Cambridge; CORUS; and the Birmingham School of Jewellery.
(3) Disclaimer: The recommendations contained in this book are of a general nature and should be customised to suit the project under consideration. Although every care has been taken in assembling this document no liability for negligence or otherwise can be accepted by the editor, authors or the publisher.
(4) In the text, considerable use has been made of acronyms (e.g. ICE = Institution of Civil Engineers). These are explained in full in the Appendix with, where appropriate, the addresses of useful websites.

Acknowledgements

This book would not have been possible without the professional assistance of many people. In particular I would like to pay tribute to my fellow authors James Douglas (Strathclyde University) and Richard Pratley (Consultant Architect), to Lawrance Hurst (Consultant to Hurst Peirce + Malcolm) for his Foreword, to Bill Black (Drivers Jonas), my long time friend and colleague, for much general advice and his thoughtful and critical review of the text, to Linsey Gullon at Whittles Publishing and to Dr Keith Whittles my publisher.

At the Institution of Structural Engineers, thanks go to Dr Susan Doran (Technical Director); Robert Thomas (Librarian) and Kathy Stansfield (Editor of *The Structural Engineer*). Also to those from whose learned papers I have used illustrations, including David Dibb-Fuller (Giffords), Barry Mawson (Gwent Consultancy), Steve Evans (Peel & Fowler), Dr Jack Chapman (Consultant and my old boss), David Yeoell (Westminster City Council), Leigh Birch (Elliott & Brown) and Professor I A MacLeod (Strathclyde University).

At the Institution of Civil Engineers thanks to Michael Chrimes, Rose Marney and others; to Robert Gerrard (Consultant), thanks for helpful advice on NEC and other contracts; to Matt Neave (Drivers Jonas) for a review of Chapter 4 and for advice on EU regulations with regard to tendering for public funded work; to George Charalambous and Colin Smart (CORUS) for assistance with information concerning steel (including weathering steel) and illustrations; to Michael Johnson (Costain), a former Wimpey colleague, thanks for helpful advice in the early stages of the project; to Clive Cockerton (Consultant), thanks for suggestions on topics and other assistance on Chapter 5; to Jamie Cant (NHBC), thanks for general advice from his organisation; to Chris Shaw (Consultant) for his help with considerations concerning cover to reinforced concrete and illustrations; to Kenneth MacAlpine for assistance with illustrations; to my wife Maureen for some photographic images and other helpful encouragement.

There may well be others, un-named, to whom I also owe a further debt of gratitude who have given their time and expertise to help me along the way. My apologies if I have failed to mention them by name.

David Doran

1 Introduction

As conservation *and* re-use *enter the mainstream of practice, young engineers discover that little or nothing in their education has prepared them for this revolution. The past has caught up with us without our recognising it.*

– Robert Bowles and Robert Thorne—*A hundred years of Structural Engineering*, 2008

1.1 General

It is generally accepted that approximately 50% of construction work involves repair and refurbishment (figures from the Building Cost Information Service, BCIS). Recent estimates have put the total value of construction at £80bn per annum, so the value of refurbishment must be in the order of £40bn. More surprisingly, it has recently been stated that more than 30% of new build contracts require remedial repairs before contract completion. This essentially practical book has been designed to meet the challenge of this type of work and is a companion to *Site Engineers Manual*, which was first published in 2004.

Michael Chrimes, the Head of Knowledge Transfer at the Institution of Civil Engineers (ICE), has emphasised in a recent paper the need for the construction profession to become more skilled at unravelling the history of existing structures and sites before work starts. He provides guidance on how that information can be made available.

All construction is risk intensive but it is the contention of the editor that repair and refurbishment may carry risks in excess of those facing a developer building on a green field site (see also Chapter 2). The book, *inter alia*, explores some of these risks and suggests ways in which they may be minimised. Chrimes, in the above mentioned article, lists the following questions to be answered before proceeding with work on existing structures:

- Where is it?
- How old is it?
- Is it the first building on the site?
- Who designed it?
- What were the original ground conditions like?
- What kind of foundations has it got?

- How was it designed?
- Who built it?
- Has its use altered?
- Has it always looked like this?
- What type of construction is it?
- What loads was it designed for?
- Is it listed or otherwise protected?

This book attempts to provide clues to answer these questions and many others.

Advice is given on sources to approach to find records of existing construction. Regrettably, in recent years, many records have been destroyed and only intensive survey investigation employing non-destructive and invasive methods can reveal an accurate picture of the construction and condition of a structure. The penalty for not so doing can be penal. In a recently reported case £30,000 of additional cost was incurred for repairs to undisclosed, defective roof timbers. This amounted to 30% of the original budget for the project.

The construction industry is complex, fragmented and sometimes inefficient. It has recently been stated that on some projects 40% of man hours and 20% of materials are wasted (Simon Murray, Osborne). Many attempts have been made to improve performance and a number of initiatives have taken place to remedy this situation. Some of these are the subject of the reports *Constructing the team* and *Rethinking construction: The report of the construction task force.*

Rethinking construction highlights the following initiatives:

Engineering best practice:
- integrated teams
- innovation
- managing risk
- health and safety
- sustainability

Delivering best service:
- key performance indicators
- managing time and change
- adding value
- inputting to the cost plan

Working together:
- contributing to the whole process
- a partnering approach
- providing new and additional values
- contributing best value

Globally, concerns have been raised about the depletion of resources due to waste and overproduction. These concerns are also highlighted in *Rethinking Construction* and have been addressed more fully in the report *Building for a sustainable future: Construction without depletion.*

The drivers for sustainable construction include:

- the European Directive on the energy performance of buildings (Part L Building Regulations);
- the Landfill Directive;
- the Construction Products Directive;
- fiscal measures and taxes (for example landfill tax, climate changes levy, primary aggregates levy);
- national and regional planning policies;
- the Code for Sustainable Building; and
- the Secure and Sustainable Buildings Act.

At the time of writing the Department for Communities and Local Government (DCLG; formerly the Office of the Deputy Prime Minister, the ODPM) issued for comment a draft *Code for Sustainable Housing*. It would appear that final publication has been delayed because of adverse comments received from practitioners.

Currently many schemes for new-build and refurbishment are subjected to an environmental assessment. This usually takes place by request of the client but the received wisdom suggests that such requirements will soon become mandatory. At least two such formal procedures exist, the Building Research Establishment Environmental Assessment Method (BREEAM) and EcoHomes, the latter being for housing. BREEAM requires a commitment in a number of areas which include:

- management: overall policy, site management via the Construction Confederation Considerate Constructors Scheme and procedural issues;
- health and wellbeing: both internal and external issues affecting the occupants' health;
- energy efficiency including operational energy and carbon dioxide;
- transport: carbon dioxide and location related factors;
- water consumption and efficiency;
- materials: environmental implications and life cycle impact;
- land use regarding green-field and brown-field sites;
- ecology including enhancement of the site as well as ecological value conservation; and
- pollution of air, land and water.

There is also growing support from architects and other like-minded professionals to encourage the use of *The Green Guide to Housing Specification* (Anderson, Howard, 2000). Now in its third edition, this slim publication contains over 150 specifications commonly used in housing. These include typical wall, roof, floor and other constructions listed against a simple environmental scale running from A (good) to C (poor). This guide enables practitioners to select materials and components on their perceived overall performance in the design life of the building (see Section 1.3).

Other initiatives have considered the cost of construction in terms of whole life costing where costs in addition to construction costs are considered. This 'cradle to grave' approach considers such issues as land cost, land reclamation, running costs (heating, lighting etc.), maintenance, change of use, upgrading to match new legislation, embodied energy in manufactured items, transport costs, demolition and re-use. Such considerations demand that clients and developers look carefully at the options of new-build or refurbishment. Although Central Government (in particular in the housing sector) has hinted that new-build may be preferable to refurbishment, it has been shown that in many cases refurbishment is the preferred option. A recent example of this was the success of refurbishing Camden Mill into the Bath Head Office of Buro Happold, as described by Richard Harris.

A well publicised case of the merit of refurbishment is that of Tinsley viaduct which failed, when re-assessed, to meet the requirement for use by 40-ton lorries. The alternatives were to close the M1 motorway and the A631 trunk road and build a new bridge or to refurbish and upgrade the existing structure. A replacement bridge was estimated to cost £200m together with associated costs of congestion rated at £1400m spread over a 2–3 year construction period. The decision was made to upgrade the bridge involving a complex strengthening process whilst keeping the road open to traffic apart from very short night time closures. The cost was £80m showing a saving of £1400m (Long et al., 2007).

Although it is essential that all these new ideas are absorbed into future construction work, there are no substitutes for accurate surveys, well prepared and timely documentation, good planning and construction carried out by skilled personnel.

As with new-build construction, careful consideration should be given to the use of off-site and prefabricated assemblies.

Recycling and sustainability need to be carefully considered. The World Commission on Environment and Development in 1987 defined Sustainable Development as 'Development which meets the needs of the present without compromising the ability of future generations to meet their own needs'. It has been stated that the earth's resources are being used up at such a rate that we need three planets the size of the earth to sustain the present rate of growth. Michael Dickson (the recent President of The Institution of Structural Engineers) has highlighted the four Rs of Sustainability as Reduce, Re-use, Refurbish and Recycle.

The steel industry claims that 94% of all steel construction products are either reused or recycled when buildings are demolished. Experimental work continues on recycling asphalt for reuse; old vehicle tyres and glass bottle derivatives are experimentally being used for road surfacing.

Running in parallel with the requirement to reduce energy is the need to reduce carbon-related emissions because of their effect on climate. The total annual sum of carbon emissions in the UK is in excess of 150 million tonnes (Mt) of which a large proportion is building or structure-related (Stansfield, 2006 and Harris, 2006). Practitioners need to recognise this and act accordingly.

The received wisdom is that the voluntary best practice techniques of today (for example the use of environmental assessments, BRE BREEAM and EcoHomes management systems) will become mandatory in the near future so it is prudent to plan refurbishment schemes with those aims in view. *The Green Guide to Specification* is also gaining a good reputation for the way forward.

In the light of initiatives for recycling it is disappointing, at the time of writing, to hear that more than 2 Mt of materials per annum of recycled UK materials finish up on landfill sites in China, India and elsewhere. It is stated that the construction industry produces 70 Mt of waste each year and of that 13 Mt are material delivered to site but never used (Stansfield, 2006 and Harris, 2006).

It has been stated that a special type of operative may be required for refurbishment work. This requires a training input from the industry, particularly in the field of apprenticeships. Perhaps a requirement is for a bolt-on segment to a regular apprenticeship to deal specifically with skills appropriate to refurbishment and repair. Although there has been something of a revival in apprentice training, more needs to be done. Contact with the Construction Industry Training Board (CITB) can assist in this process. It is also relevant to note that a partnership has been concluded between the Construction Industry Council (CIC), CITB Construction Skills and CITB Northern Ireland to form 'ConstructionSkills', the key goals of which are:

- reducing skills gaps and shortages;
- improving performance;
- boosting skills and productivity; and
- improving learning supply.

It is also possible that The City and Guilds Institute may introduce a National Vocational Qualification (NVQ) in this area.

Refurbishment and repair can be a dangerous business. The collapse of at least one house in the North of England has occurred because the builder misunderstood the load-bearing attributes of an internal brick partition. The wholesale removal of cross walls from terraces of housing may also introduce the ingredients for the domino effect and seriously reduce the robustness of the original construction.

1.2 Reasons for refurbishment and repair

These are many and varied and include:

- change of use (including change of loading requirements)
- dilapidation
- fire damage
- explosion damage

- vehicle, (ship, plane and train) impact
- upgrading to meet new legislation (for example The Disability Discrimination Act, new insulation requirements etc)
- settlement or heave of foundations
- structural inadequacy (this may result from collision damage, long term deterioration in, for example, marine structures, reinforcement corrosion, corrosion of structural steelwork or inadequate original design or detailing)
- security requirements
- new plant or mechanical services requirements
- additional client requirements (for example enhancements from marketing demands)
- maximisation of space to let
- expiry of lease

1.3 Design life

One of the most difficult questions asked of his design team by a client will be 'how long will my refurbished building last?' The response to this requires a great deal of thought and can only be given accompanied by a number of qualifications! BRE have recently developed a methodology for new structures some of which can be applied to other types of structure. BS 8110 does give some predictions for new structures based on material specification, concrete cover, and a presumption of good workmanship. For existing structures and using BS 8110 advice it may be possible to make a reasoned estimate of the residual design life for concrete structures. Steen Rostam (1990) and others have suggested that the depth of penetration of carbonation in a reinforced concrete structure is a guide to how far through the life cycle the structure has progressed. In some cases it may be possible to prolong the life of a concrete structure by applying a coating to the external surfaces.

When considering design life of concrete structures it is worth adopting the discipline suggested by WR de Sitter at a CEB/RILEM workshop in 1983 at which he suggested the following four phases of construction as:

A: design, construction and curing period
B: initiation processes underway, no propagation of damage has begun yet
C: propagating deterioration has just begun
D: advanced state of propagation with extensive damage occurring

These phases and the associated costs are based on experience of the so-called Law of Fives which states:

One dollar spent on Phase A equals five dollars in Phase B, equals twenty-five dollars in Phase C, equals one-hundred and twenty-five dollars in Phase D!

Several organisations have attributed desirable service lives to structures of which a composite sample is detailed below:

- temporary buildings – up to 10 years
- most industrial buildings – 30 years
- offshore structures – 35 years
- structures designed according to international and national codes – 50 years
- new housing – 60 years
- structures designed according to international and national codes – 50 years
- bridges, tunnels, harbours – 100 years (120 years in the UK)
- storm surge barriers – 200 years
- Messina Strait Bridge – 300 years

Any inconsistencies in these figures indicate the difficulty in obtaining national and international agreement on these matters.

The British Standards Institution (BSI) for a number of years has been collaborating with the International Organisation for Standardisation (ISO) to produce ISO 15686 *Buildings and constructed assets* to deal with complex issues of durability. When complete it is hoped that under Vienna Convention arrangements that this standard will be adopted by Europe. In the meantime BS 7543:1992 *Guide to durability of buildings and building elements, products and components* is still available.

1.4 Forensic investigations

When carrying out forensic investigation prior to executing repairs things may not be as straightforward as they appear. In a paper to the Royal Society, Dr Bill Allen of Bickerdike Allen Partners wrote:

> In the hot dry summer of 1976, when no drop of rain fell in southern England for many weeks, our firm received an unexpected request to investigate what was perceived to be leakage through a flat roof of a large hotel. The support deck was of precast concrete slabs carrying a vapour barrier, a considerable depth of a coarse granular insulant, and an asphalt finish of good quality. No fault could be found in the asphalt yet there was 30-40mm of water lying on the vapour barrier in places.
>
> The leakage down into the rooms was in the form of dark brown water dripping through the joints between pre-castings and, remarkably, it took place only in the sunshine, so what was going on here? The fact that the water could easily get through the so-called vapour barrier argued that it was a low-grade product, and another fact, that the water was brown, suggested that it was alkaline, for alkaline moisture can react with low-grade bituminous materials such, perhaps, as are found in a low-grade vapour barrier and acquire brownishness. If the 30 or 40 mm of water had been rain it would normally be acidic. So how would this water have become alkaline?
>
> For a reason, unknown at the time, the air in the rooms was very humid in spite of air conditioning and this suggested that the water vapour was going up through the joints between the pre-castings, becoming alkaline by contact with the mortar, and then getting through the vapour barrier by whatever means the leakage water used when coming down. Once inside the insulation zone it would condense on the underside of the asphalt when this cooled off

at night and dribble down through the insulant to accumulate as liquid water on the vapour barrier.

But what was the force moving the vapour up through the joints and the water downward through the joints in the first place? The vapour forming the room humidity is of course a gas and exerts independent pressure, but this hardly seemed sufficient and a better clue seemed to lie in the fact that the dripping only took place in sunshine, arguing that the solar warming of the air in the insulation zone could raise its pressure sufficiently to draw vapour up through the joints at night. Thus we arrived at [the] perception of a cyclic pressure mechanism which we described as thermal pumping.

But there was a final question; why was the air in the room so humid. This proved to be an accident of installation. The intention had been, as, usual, to admit a proportion of fresh air to the re-circulation and this had not been done. What was re-circulating therefore was simply the accumulating breath of the hotel's patrons.

It is therefore imperative that a thorough, appropriate and professional investigation is carried out before embarking on any refurbishment or repair. On more general issues, it is worth noting that obsolescent buildings may, for example, require extra storey height to accommodate additional ventilation or under-floor services. These provisions may be difficult or impossible to achieve.

1.5 Learning from the past

The construction industry needs to overcome the collective amnesia that inhibits the need to learn from the past (discussed in detail in Chapter 5 of this book). Repetitive problems include:

- Brick walls built without movement joints. Joints, where they exist, are routinely over-capped by coping stones in parapet walls (see Figs 1.1 and 1.2).
- Thermal imaging frequently showing up a deficiency of wall ties in cavity masonry construction.
- Multi-storey buildings constructed out-of-plumb causing support problems for cladding at the higher levels. Cladding construction is sometimes commenced without a dimensional frame survey to detect out of tolerance inaccuracies.
- Housing estates constructed on sites devoid of adequate site investigation. Sub-surface mining not being properly assessed resulting in catastrophic settlements or sensitive clays not being adequately assessed resulting in the provision of inadequate foundations
- Reinforced concrete produced with inadequate cover to reinforcement causing premature corrosion. It has recently been estimated that the cost of remedial measures to correct this defect alone runs to £550m per annum.
- The correct balance between heating and ventilation not always being achieved.

Figure 1.1 Long length of brickwork without movement joints.

Figure 1.2 Cracking and displacement due to lack of joints.

- Cavity wall insulation sometimes being installed in ways that are detrimental to the basic principles of cavity construction.
- Scaffolds and tower cranes continuing to collapse. Tower cranes appear to be at their most vulnerable when the vertical tower is being raised
- In other, unconnected incidents, in 2007, HSE put a temporary stop on the use of cranes from one supplier until all had been checked by competent engineers.

As a result, construction sometimes deservedly gets a bad name and litigation lawyers continue to prosper.

1.6 Europe

Practitioners need to be aware of the increasing number of restrictions imposed on the industry by the European Union (EU). In particular these include:

- Compliance with the Construction Products Directive (CPD) (see Chapter 6).
- Recognition that tenders in excess of a certain threshold need to be open to EU competition (see Chapter 4).
- The gradual move to embrace European structural and other standards. The current plan is to withdraw British Standards and Codes of Practice by 2010 and replace them with Eurocodes. At the time of writing the list of structural Eurocodes was as shown below:
  - BS EN 1990 Basis of structural design
  - BS EN 1991 Eurocode 1: Actions on structures
  - BS EN 1992 Eurocode 2: Concrete
  - BS EN 1993 Eurocode 3: Steel
  - BS EN 1994 Eurocode 4: Composite
  - BS EN 1995 Eurocode 5: Timber
  - BS EN 1996 Eurocode 6: Masonry
  - BS EN 1997 Eurocode 7: Geotechnical design
  - BS EN 1998 Eurocode 8: Seismic design
  - BS EN 1999 Eurocode 9: Aluminium

These codes will become virtually mandatory except that, as is the practice with UK standards, other methods will be allowed if precise criteria of justification are met.

Each of these codes will be accompanied by a National Annexe (NA) which will set out specific parameters that apply only to the country in which they appear. NAs will reflect local environmental and other nation-specific factors. (Note: Recent discussions with the EU have led to uncertainty over the extent of adoption, in the UK, of metric measurements.)

Although not universally accepted it is claimed by some that the benefits of using Eurocodes are that:

- They are claimed to be the most technically advanced in the world.
- Eurocode 2 (Concrete) should result in more economical structures than BS 8110.
- They are logical and organised to avoid repetition.
- Eurocode 2 is less restrictive but more extensive than existing codes.
- The use of eurocodes will provide more opportunity for designers to work throughout Europe.
- In Europe all public works must allow Eurocodes to be used.

1.7 Energy conservation

At the time of writing the cost of energy was rising rapidly. In the UK these rises were particularly acute due to the depletion of oil and gas reserves in the North Sea and the consequent heavy reliance on sometimes volatile foreign supply sources. Coupled with this cost issue goes the need to reduce carbon dioxide emissions which contribute to global warming. It follows that the refurbishment of buildings must respond to this challenge in a number of ways including:

- Providing good insulation (basically complying with or exceeding the requirements of Part L of the Building Regulations).
- Using products of optimum embodied energy (embodied energy is defined as the amount of energy consumed in manufacture, measured in kilowatts per hour per cubic metre (kWh/m^3). Recent examples (two of which are shown below) have drawn attention to the difference in embodied energy between commonly used products and low energy alternatives:
  - Mass-produced clay bricks from factories in the Midlands. Bricks require massive amounts of energy to fire them (1500 kWh/m^3) and more energy is absorbed during transportation to sites. Alternatively, locally produced Fletton bricks whose embedded impurities aid the firing process (300 kWh/m^3) or reclaimed bricks from demolition sites *may* be a cheaper and more energy efficient option.
  - PVC framed windows: PVC produces large amounts of toxic waste during production and requires high amounts of embodied energy (47,000 kWh/m^3). Alternatively, timber-framed double-glazed units made with timber from Forest Stewardship Council (FSC) certified forests may prove a ecologically better choice.

The book *The Ecology of Building Materials* (Berge, 2009) helpfully lists the various stages of energy use in construction as follows:

- Energy consumption during the manufacture of building materials (usually about 80% of the total energy input):
  - the direct energy consumption in extraction of raw materials and the production processes

  – secondary consumption in the manufacturing process
  – energy in transport of the necessary raw and processed materials (a
    Norwegian study has assessed the energy consumption (in MJ/ton/km)
    to vary between 0.2 and 1.6 for electric rail transport as opposed to die-
    sel road transport respectively)
- Energy consumption during building, use and demolition (usually about
  20% of the total energy input):
  – energy consumption for the transport of manufactured products
  – energy consumption on the building site
  – energy consumption during maintenance
  – energy consumption of dismantling or removal of materials during
    demolition

Obviously these are basic guidelines and proportions will vary according to materials used and the possible degree of recycling of demolition materials. Care must
be taken in assessing embodied energy: for example, although making of steel is
an efficient process it may have been imported from India or South America thus
adding a considerable transport burden to the energy equation.

In a groundbreaking paper David Collings (of Consultants Benaim) has suggested
that for bridges the embodied energy/carbon dioxide balance may be different from
that when considering buildings (Collings 2006). In so doing he has formulated the
following rules for use as guidelines when considering such matters:

- The initial environmental burden of a bridge will be approximately
  proportional to its cost. A bridge with low material content and repetitive
  construction technique is likely to have lower embodied energy and to
  minimise CO_2 emission.
- The ongoing environmental burden will be approximately proportional to
  the amount of maintenance required. A bridge requiring regular repainting
  or replacement of joints and bearings is likely to have increased energy use
  and CO_2 emissions in comparison to one without these elements. (Here
  it needs to be remembered that carrying out major works may require a
  bridge closure and considerable traffic management for that period.)
- There is little difference in overall environmental burden between steel and
  concrete structures.
- The added energy and CO_2 emissions over the life of the structure, during
  repair and maintenance, will be larger that the initial values generated
  during construction. As bridge structures rarely require heating, air
  conditioning or other services associated with occupation, this additional
  burden is likely to be less than for a building.

In a recent study Dr Fergal Kelly of Peter Brett Associates (PBA), during a project
for Oxford University, arrived at the conclusion that structural steel rather than
reinforced concrete produced significant sustainability benefits in terms of energy

saving and reduced CO_2. It should be stressed, however, that energy calculations must be computed with care and include all aspects, such as transportation, to ensure accurate results.

In all these considerations it is imperative that a whole life cost analysis is made to assist in sensible decision making.

1.8 Unnecessary repairs

Dr Chris Burgoyne of Cambridge University, in a paper to the Institution of Structural Engineers in 2004, posed the question: 'Are structures being repaired unnecessarily?' (Burgoyne 2004). He traces the history of a hypothetical reinforced concrete slab simply supported on four sides originally designed in 1970. This is then re-analysed today using contemporary methods and found wanting although it has performed adequately for over 30 years! The lesson from this is not to dismiss too easily confidence of use in the simple methods of design that have stood the test of time.

1.9 Safety of historic structures

Many historic structures need to be analysed in ways that need a great deal of experience and imagination.

In a recent paper David Yeomans (Yeomans 2006), the secretary of the International Council on Monuments and Sites (ICOMOS), has stated that 'Design codes drawn up for new construction are incompatible with historic buildings-and we need to develop different approaches equally acceptable to safety authorities'. Under the guidance of Professor Giorgio Croci, the International Scientific Committee for the Analysis and Restoration of Structures of Architectural Heritage (ISCARSAH) have produced a set of recommendations which have now been published as an ICOMOS charter together with guidelines for the benefit of practitioners.

Yeomans concludes his paper by recommending the need for:

- training;
- proper management for conservation work; and
- further studies of the behaviour of historic structures that can aid qualitative approaches to safety.

Poul Beckmann and Robert Bowles in their book *Structural Aspects of Building Conservation* (Beckmann and Bowles, 2004) have taken a more pragmatic approach and suggest a return to basic principles. In an illuminating chapter refreshingly free from jargon and mathematics they deal with a number of basic concepts including:

- equilibrium of external forces
- equilibrium of internal forces
- triangle of forces

- behaviour under tension
- behaviour under compression
- behaviour under bending
- behaviour under shear
- buckling
- behaviour due to temperature and moisture fluctuations
- the effect of restraints
- stability and robustness

(To this list it would also be appropriate to add structural redundancy.)

In this connection Poul Beckmann is alleged to have said 'in Roman times they sacrificed a slave to ensure that a building stood up. Nowadays we do calculations!'

1.10 Conservation

It is important that those dealing with the conservation of historical works and sites are sufficiently skilled to carry out those tasks. In this connection the ICE and Institution of Structural Engineers (IStructE) have jointly arranged to register those engineers who are suitably qualified. The scheme is known as the Conservation Accreditation Register for Engineers (CARE). Accreditation rests on the following principles:

- accreditation of individuals;
- eligibility of all suitably qualified and experienced professional engineers to gain accreditation;
- assessment by suitably experienced assessors; and
- assessment of verifiable records of case studies.

Registration is for five years after which it can be renewed by the submission of evidence to the CARE Panel that candidates have maintained and enlarged their skills by further relevant experience and training.

ISCARAH have produced a set of recommendations which have now been published as the ICOMOS Charter, Principles for the Analysis, Conservation and Structural Restoration of Architectural Heritage, together with guidelines for the benefit of practitioners.

1.11 Final observations

This book should be of interest to developers, architects, surveyors, civil and structural engineers, students (and their lecturers) and contractors. It aims to guide practitioners through an increasingly complex maze of legal and other demands to a satisfactorily completed project. Quality, safety, completion on time (and to budget) and profitability are the essential ingredients of a successful contract. This book should help to achieve that justifiable aim.

Bibliography and further reference

Allinson, K. (Ed.) 2006. *London's Contemporary Architecture,* 4th edn., Architectural Press, Oxford.

Anderson, J. and Howard, N. 2000. *The green guide to housing specification,* Building Research Establishment, London.

Armstrong, J.H., 1987. Refurbishment and renovation: the city centre. In: Sandberg A. (Ed.) *Rehabilitation and renovation,* Proceedings of the 1987 Henderson Colloquium, organised by the British Group of the IABSE, London.

BCIS. 2005. *The economic significance of maintenance,* Building Cost Information Service, London.

BCSA. 2008. *Sustainable steel construction,* BCSA Steel Industry Guidance Note S929 10/2008, British Constitutional Steel Association, London.

Beckmann, P. and Bowles R. 2004. *Structural aspects of building conservation,* 2nd edn., Elsevier, Oxford.

Berge, B. 2009. *Ecology of building materials,* 2nd edn., Architectural Press, Oxford.

Biggs, W. 1987. Repair and maintenance of buildings. In: Sandberg A. (Ed.) *Rehabilitation and renovation,* Proceedings of the 1987 Henderson Colloquium, organised by the British Group of the IABSE, London.

Bowles R. and Thorne R. 2008. Conservation, refurbishment and re-use of buildings. *The Structural Engineer,* 86(14), IStructE, London.

BRE. 1991. *Structural appraisal of existing buildings for change of use,* Building Research Establishment, London.

Burgoyne, C. 2004. Are structures being repaired unnecessarily? *The Structural Engineer,* 82(1), IStructE, London.

Burns, J.G. 1990. Design life of buildings: client expectations. In: Somerville, G. (Ed.) *The design life of structures,* Proceedings of the 1990 Henderson Colloquium, organised by the British Group of the IABSE, London.

Campbell, P. (Ed.) 2001. *Learning from construction failures: applied forensic engineering,* Whittles Publishing, Scotland.

Chapman, J.C. 1998. Collapse of the Ramsgate walkway, *The Structural Engineer,* 76(1), IStructE, London.

Chapman, J.C. 2000. Collapse of the Ramsgate walkway (Discussion), *The Structural Engineer* 78(4), IStructE, London.

Chrimes, M. 2006. Historical research: a guide for civil engineers, *Proceedings of ICE, Civil Engineering,* 159(1), Thomas Telford, London.

Coventry, S., Shorter, B. and Kingsley, M. 2001. *Demonstrating waste minimisation benefits in construction,* CIRIA, London.

Collings, D. 2006. An environmental comparison of bridge forms, *Proceedings of ICE, Bridge Engineering,* 161(4), Thomas Telford, London.

de Sitter, W.R. 1984. Costs for a service life estimation: The law of fives. In: *Durability of Concrete Structures,* Workshop Report, CEB-RILEM Workshop 18–20 May 1983, Copenhagen.

Doran, D.K. (Ed.). 2009. *Site Engineers Manual,* 2nd edn.,Whittles Publishing, Scotland.

Eaton, K.J. 1998. Towards sustainable construction. In: Pickett, A. (Ed.) *Structures beyond 2000,* Proceedings of the 1998 Henderson Colloquium, organised by the British Group of the IABSE, London.

Egan, J. 1998. *Rethinking construction: the report of the construction task force*, Egan Report, Department of the Environment Transport and Regions Construction Task Force, London.

English Heritage. 1994. *Office floor loading in historic buildings*, English Heritage, Swindon. Information leaflet (4pp).

Gordon, J.E. 1981. *Structures: or why things don't fall down*, Pelican, London.

Hammond, R.E. 1967. *Structural failures in civil engineering works*, The Concrete Society, Surrey.

Harris, M. 2006. Working together to face the low carbon economy, *The Structural Engineer*, 84(9), IStructE, London.

Haseltine, B.A. 1981. Energy used in manufacture of materials. In: Cusens, A. (Ed.) *Materials in structures*, Proceedings of the 1981 Henderson Colloquium, organised by the British Group of the IABSE, London.

Heyman, J. 1967. Westminster Hall roof. *Proceedings of the ICE*, 37(1), Thomas Telford, London.

Heyman, J. 1996. *Arches vaults and buttresses: masonry structures and their engineering*, Variorum, Aldershot.

IStructE. 1999. *Building for a sustainable future: construction without depletion*, IStructE, London.

Jordan, G. W. 1990. Prediction of design life of structures in the nuclear reprocessing industry. In: Somerville, G. (Ed.) *The design life of structures*, Proceedings of the 1990 Henderson Colloquium, organised by the British Group of the IABSE, London.

BCSA. 2008. *Sustainable steel construction*, BCSA Steel Industry Guidance Note SN29 10/2008, BCSA, Westminster.

Latham, M. 1994. *Constructing the team*, Latham Report, HMSO, London.

Long, A.E., Basheer, P.A.M., Taylor, S.E., Rankin, B.G.I. and Kirkpatrick J. 2007. Sustainable bridges through innovative advances, *Proceedings of the ICE, Bridge Engineering*, 161(4), Thomas Telford, London.

Manning, D.G. 1990. Design life of concrete highway structures: the North American scene. In: Somerville, G. (Ed.) *The design life of structures*, Proceedings of the 1990 Henderson Colloquium, organised by the British Group of the IABSE, London.

Menzies, J. 1990. Design life and populations of building structures. In: Somerville, G. (Ed.) *The design life of structures*, Proceedings of the 1990 Henderson Colloquium, organised by the British Group of the IABSE, London.

Ogle, M.H. 1990. Design life of welded structures. In: Somerville, G. (Ed.) *The design life of structures*, Proceedings of the 1990 Henderson Colloquium, organised by the British Group of the IABSE, London.

Paterson J. and Perry, P., 2006. A systematic approach to refurbishment, *The Structural Engineer*, 80(9), IStructE, London.

Piesold, D.D.A. 1991. *Civil Engineering Practice: engineering success by analysis of failure*, McGraw-Hill, Maidenhead.

Price, S. 1996. Cantilevered staircases, *Architectural Research Quarterly* 1(3), Cambridge University Press, Cambridge.

Rostam, S. 1990. The European approach to design life. In: Somerville, G. (Ed.) *The design life of structures*, Proceedings of the 1990 Henderson Colloquium, organised by the British Group of the IABSE, London.

Safier, A. 1987. Materials and design life. In: Sandberg A. (Ed.) *Rehabilitation and renovation,* Proceedings of the 1987 Henderson Colloquium, organised by the British Group of the IABSE, London.

SCI. 2007. *Sustainable Construction in Steel – Information Sheet 2: BREEAM Environmental assessments,* Steel Construction Institute, Ascot.

Somerville, G. 1990. Some final reflections on design life. In: Somerville, G. (Ed.) *The design life of structures,* Proceedings of the 1990 Henderson Colloquium, organised by the British Group of the IABSE, London.

Somerville, G. (Ed.) 1992. *The design life of structures,* Blackie, Glasgow.

Stansfield, K. 2006. Working together to face the low carbon economy, *The Structural Engineer,* 84(9), IStructE, London.

Stillman, J. 1990. Design life and the new code. In: Somerville, G. (Ed.) *The design life of structures,* Proceedings of the 1990 Henderson Colloquium, organised by the British Group of the IABSE, London.

Yeomans, D. 2006. The safety of historic structures, *The Structural Engineer,* 84(6), IStructE, London.

Yu, C.W. and Bull, J.W. 2006. *Durability of materials and structures: in building and civil engineering,* Whittles Publishing, Scotland.

2 Risks

2.1 Preamble

All construction projects incur risks. However, because of their unique or unusual nature refurbishment schemes often entail just as many, if not more, risks than comparably sized new-build projects. This is because the level of uncertainty (see below) and the degree of constraints in the former type of construction work are often greater than the latter. For example, the main constraints affecting construction projects, especially those involving refurbishment are:

- Financial: funds for refurbishment and other similar work might be more restricted as part of an organisation's cost-cutting exercise.
- Spatial: existing building dimensions already fixed, leaving little or no room for expansion.
- Temporal: work is usually expected to be completed within a shorter time span.
- Legal: full code compliance might be more difficult to achieve because of physical restraints, such as type of existing construction, lack of space, etc.
- Personnel: refurbishment is labour intensive, particularly with regard to skilled workers.

Guidance is given in this chapter on how to minimise/manage risk on refurbishment projects and how to establish a 'risk profile'. The scope of a risk profile should cover risks arising from the client, contractor, design, contract, cost and programme. Risk allocation should identify those best able to mitigate risk.

2.2 Nature of risk

Risk generally has a negative connotation in that it usually involves some form of financial loss. It has also been defined as the combination of the possibility of an event and its consequence (BSI 2002). Alternatively, risk can be seen as the chance that an actual outcome will deviate from that forecast or intended.

Fatality is the severest type of risk for human beings. The relative risk of death in UK by activity is shown in Table 2.1. It is expressed in terms of the fatality accident rate (FAR), which is the risk of death per 100 million hours of exposure to the activity.

Table 2.1 Relative risk of death in the UK by activity (based on Hambly and Hambly, 1994).

Activity	Risk of death x 10^{-8} h: FAR[†]
1. Plague in London in 1665	15000
2. Rock climbing, while on rock face	4000
3. Fireman in London air-raids 1940	1000
4. Travel by helicopter	500
5. Travel by motorcycle and moped	300
6. Police officer in Northern Ireland, average	70
7. Construction, high-rise erectors	70
8. 'Tolerable' limit 1 in 1000/yr at work	50
9. Smoking	40
10. Walking beside a road	20
11. Offshore oil and gas extraction	20
12. Travel by air	15
13. Travel by car	15
14. Coal mines	8
15. Average man in 30s from accidents	8
16. Average man in 30s from diseases	8
17. Travel by train	5
18. Construction, average	5
19. Metal manufacturing	4
20. Travel by bus	1
21. Accident at home, able-bodied	1
22. 'Tolerable' limit 1 in 10000/yr near major hazard	1
23. Radon gas natural radiation 'action level'	5
24. Radon gas natural radiation, UK level	0.1
25. 'Tolerable' limit 1 in 100000/year near nuclear plant	0.1
26. Terrorist bomb in London street	0.1
27. Buildings fall down	0.002

† FAR = fatality accident rate: the risk of death per 100 million hours of exposure to the activity.

Note: All figures are approximate and depend on a number of assumptions as detailed in Hambly and Hambly (1994).

Closely related to risk is uncertainty. The main difference between them is that the former is something that is considered to be reasonably objective in nature and thus quantifiable, whilst the latter is more subjective but generally unquantifiable (Douglas 2006). Risk always involves an element of uncertainty, but uncertainty does not always involve risk.

The differences between risk and uncertainty are illustrated in Figs. 2.1 and 2.2. Fig. 2.1 can be contrasted with the risk-uncertainty spectrum shown in Fig. 2.2.

	Risk	Certainty
⬅	SPECTRUM OF RISK	➡
Unknown Unknowns	Known Unknowns	Known Knowns
No information (No 'as built' drawings)	Partial information (Some original drawings)	Complete information (Full set of 'as built' drawings and specification)

Figure 2.1 The basic risk spectrum (Based on Bowles and Kelly 2005).

Risk	**Uncertainty**
Quantifiable	Non-Quantifiable
Statistical assessment	Subjective Probability
'Hard' data	Informed Opinion

Figure 2.2 The risk-uncertainty spectrum (Based on Douglas 2006; Raftery 1993).

The degree of 'uniqueness' of a construction project strongly influences the degree of associate risk and uncertainty. This is determined by the homogeneity and heterogeneity of a refurbishment scheme.

There is less uncertainty and thus lower risk associated with the homogeneous characteristics of a building project. These are summarised as follows:

- Elements of a building: roof, walls, floors, etc.
- Building components: doors, windows, etc.

- Construction materials: bricks, mortar, concrete, steel, timber, etc.
- Site operations: materials deliveries, lifting, installing, etc.
- Management structure and style: project manager, site agent, and foremen.

Heterogeneity, on the other hand, is a more distinguishing characteristic of all construction projects. It is this which creates the conditions of greater uncertainty about the outcome of events or situations on sites. In consequence there is a higher degree of risk with construction work. According to Bowles and Kelly (2005) it results in a range of project differences such as:

- Building and site conditions:
- With perhaps the exception of some housing estates no two buildings have exactly the same exposure conditions, performance levels, access provisions, etc. Every property is therefore unique in some way.
- Element specifications:
- The actual specification and design of new (e.g. replacement) and existing elements is wide and varied. This also depends on whether a prescriptive (i.e. traditional or recipe) specification or performance (i.e. modern or functional) specification is being used.
- Management structure and style:
- Personnel compositions and levels are never the same, and the forms of contract used are often adapted with variations and amended clauses, all of which can militate against standard forms of contract.
- Sources of labour:
- The fragmented nature of the building industry's labour and its discontinuous supply chain makes forecasting and planning difficult. However, this is often offset by the fact that, despite being more labour intensive, refurbishment contracts may have fewer operatives on site at any one time than a comparable new-build scheme. Also, the influx of foreign workers (e.g. East Europeans) can create language difficulties when dealing with them on site.

Risk, though, is not just the consequence of a particular unidentifiable event. It is also the result of the 'condition' or 'set of circumstances' that exists in the refurbishment project environment. The five main conditions/circumstances surrounding a refurbishment project that create risk (the first four as prescribed by Ward and Chapman 1999) are:

- Uncertainty about the basis of estimates or budget costs, e.g. what was the source or basis of cost data?
- Uncertainty about design and logistics, e.g. is the design based on old original drawings or on the dimensions derived from a measured survey of the building?

- Uncertainty about objectives and priorities, e.g. is the contract based on a realistic Gantt chart?
- Uncertainty about project organisation, e.g. who is in overall control/charge of the project?
- Uncertainty about the existing building's condition and performance, e.g. to what extent do the building's main elements require upgrading to rectify defects and comply with the building regulations? (However, the building regulations cannot be applied retrospectively in some cases.)

Another term related to risk is hazard. A hazard is anything that can cause harm – such as chemicals, electricity, working from ladders, etc. It can result in the risk of physical damage, injury or death. Risk is the chance, high or low, that somebody will be harmed by the hazard (HSE 2003a).

Building sites, whether they involve new build or refurbishment, are clearly hazardous places to work (Holt 2005). Later in this chapter the principal hazards and other risks associated with refurbishment are identified and typical control measures are suggested.

2.3 Risk management

Before we consider hazards, however, it would be useful to look at the risk management and the risk chain. According to the risk management section of the Health and Safety Executive's (HSE) website:

> Risk management is the process by which an organisation reaches decisions on the steps it needs to take to adequately control the risks which it generates or to which it is exposed, and by which it ensures those steps are taken.

The construction industry and its professional bodies have recognised the importance of risk management to achieving better health and safety on site. This has been prompted by the rise in accidents and fatalities in construction in recent years as reported by the HSE.

Education is one of the ways in which this trend can be reversed. For example, the Joint Board of Moderators (JBM), a body that validates the educational requirements for four leading engineering institutions, has prescribed certain guidelines that provide an excellent set of health and safety learning outcomes for all those involved in construction. These guidelines are set out in Table 2.2.

2.4 Risk chain

To put simply the risk chain can be seen as a source-event-effect process. Table 2.3 illustrates the main components of this chain.

Table 2.2 Suggested learning outcomes for health and safety risk management in built environment degree courses (based on JBM 2005).

Attitude	Ability to: appreciate the ethical view; recognise that health and safety is integral with all we do, that it is everyone's responsibility
Details in:	• Why Health and Safety is Important • Risk Management • Occupational Health
Competence	Ability to: be able to implement the basic risk assessment process
Details in:	• Risk Management • Project Examples • Health and Safety Management
Knowledge	Ability to: fulfil legal responsibility; understand the legal framework; understand the value of health and safety and its role in the construction process; recognise the influence of human behaviour; appreciate the benefits of learning from history.
Details in:	• Health and safety law • CDM Regulations 2007 • Learning from Accidents • Industry Initiatives • Health and Safety Management • Risk Assessment: Client Strategy

2.5 Risk attitude

2.5.1 General attitude

Attitudes to risk can be classified into two broad groups: general and specific. General attitude to risk is fundamental to ensuring good standards of health and safety in construction. It is a perception that should be, but often is not, uniform across the industry. The criteria for the general attitude towards risk are:

- A health and safety culture (i.e. 'the way we do things around here with respect for safety') should be promoted regularly on every construction job and at every level.
- A good health and safety record should be publicised and rewarded on site.
- Zero tolerance should be given to poor health and safety practices.
- The importance of gaining feedback on and learning from failures should be recognised.

Table 2.3 Main components of the risk process.

SOURCE ⟶	EVENT ⟶	EFFECT
Where does the risk come from?	How does the risk manifest itself?	What is its impact on the key success criteria (i.e. Time, Cost, Quality)?
Examples	Examples	Examples
• Adverse weather conditions	• Heavy rainfall leading to flooding (especially of basements)	• Damage to building fabric (Quality)
• Volatile business/ economic conditions	• Fluctuations in labour/ materials costs	• Increase in construction prices (Cost)
• Project participant (supplier, contractor, consultant, client)	• Late or wrong delivery of materials; not performing effectively	• Delay in the contract (Time)
• Dangerous, deficient or unsafe working practices	• Hot working processes, fire, explosion, gas cylinders, dropped loads	• Accident, injury, damage

2.5.2 Specific attitudes

Everyone involved in a construction will have a different specific attitude to risk, which covers types other than health and safety. Many, for example, will be risk averse, some risk neutral, and a very few risk seeking, particularly in relation to financial risks. Table 2.4 illustrates the different attitudes of the various players in the construction process according to their propensity to take risks. Those with entrepreneurial or speculative impulses (such as developers) are more likely to take risks (usually financial in nature) than those who have greater responsibility over health and safety matters (e.g. consulting engineers).

2.6 Categories of refurbishment risks

Despite the efforts through statutory requirements such as the Construction (Design and Management) Regulations (CDM Regulations) 1994 and 2007, there has been an increase in construction accidents in recent years. For example, according to the

Table 2.4 Range of risk takers (adapted from Bowles and Kelly 2005).

Risk Taking Range	Participants
HIGH ↑ ↓ LOW	• Developers • General Contractors • Specialist Contractors • M&E Contractors • Electrical Contractors • Clients • Architects • Building Surveyors • Project Managers • Quantity Surveyors • Engineering Consultants

HSE website the main causes of the fatal accidents on construction sites were:

- falling through fragile roofs and rooflights
- falling from ladders, scaffolds and other high work places
- being struck by excavators, lift trucks or dumpers
- being struck by falling loads and equipment
- being crushed by collapsing structures

The above list summarises some of the main health and safety risks in construction. The other three main categories of risk in refurbishment are: technical, management/economic and legal/commercial. Examples of each group of risk that may affect any refurbishment project, though not exhaustive, are presented in Tables 2.5, 2.6, 2.7 and 2.8.

It should be borne in mind, however, that most of the health and safety risks summarised in Table 2.5 can be minimised, if not avoided, with proper training. Also it should be noted that recent decrees by HSE require all Site Visitors to possess a Construction Skills Certification Scheme (CSCS) authorisation card.

2.7 Use of cranes in refurbishment

One main risk area when refurbishing, extending or altering buildings that is often overlooked is crane usage. See Doran (2009) for more guidance on this potentially hazardous activity.

In some ways using cranes on, over or next to existing buildings can be more dangerous than cranes on many new build schemes that operate on a clear or open

Table 2.5 Possible health and safety risks.

Example	Typical Cause
Falls from roofs, balconies, scaffolding. landings or other raised platforms	Open or inadequately secured balustrades. Defective/damaged deck.
Colliding or bumping into scaffolding or other protrusions	Unmarked and/or unprotected/unlit scaffolding or other protrusions. Poor site control of vehicles.
Collision with transport vehicles.	Excessive speed.
Struck by dropped or falling objects.	Sudden loss of attachment from cranes/trucks. Lack of toe-boards on scaffolding.
Tripping or slipping on surfaces	Irregular, contaminated, untidy or inadequately covered surfaces. Power cables on floor loose or not covered properly.
Exposures to pollutants such as high levels of noise (causing hearing damage), or inhalation of dust, smoke and other noxious substances.(causing respiratory problems).	Poor operational control of hazardous operations. Failure to wear appropriate PPE. Breaking, cutting, grinding or drilling cementitious and other particulate material.
Disturbance of or damage to asbestos – resulting in safety hazard and leaving part of structure vulnerable to fire damage.	Reckless operations where deleterious materials are present.
Standing on or being ripped by protruding nails.	Untidy working areas.Failure to wear appropriate PPE.
Crushing owing to collapse of structural elements or temporary supports in an existing wall or illicit removal of load-bearing wall.	Inadequate support or reckless formation of new opening.
Electrocution.	Contact with faulty equipment or bare wiring.
Improper handling of materials resulting in back injuries.	Lack of manual handling training.
Cuts, punctures and other injuries from hazardous working operations.	Careless use of saws, electrical drills/saws, generators and other electrical equipment, nail guns, etc.
Burns or fire spread from hot working processes.	Failure to wear appropriate PPE. Careless welding, flame cutting, laying asphalt, torching on bituminous felt, hot air paint-stripping gun, etc.
Blinding or fire spread from welding processes.	Lack of control or restrictions on work; inadequate screening.
Asphyxiation.	Working in confined space.Being exposed to noxious gases.Lack of breathing aids.
Contact with hazardous building materials (such as lime, which is caustic) and chemicals (such as ammonia).	Failure to wear appropriate PPE (e.g. gloves, etc). Non-compliance with COSHH regulations.

Table 2.6 Possible technical risks.

Example	Typical Cause
Access problems or restrictions (e.g. involving site huts and contractor's accommodation being erected on gantry at front of building, requiring approval from local authority).	Restricted site – lack of space for parking, site office accommodation and storage.
Incomplete or inaccurate information of existing structure on the available drawings - resulting in missing or incorrect dimensions on plans, elevations or sections.	Inadequate measured survey.
Hidden defects (e.g. corrosion and timber decay).	Inadequate building investigation, the full extent of defect may not be revealed until work commences.
Other unforeseen defects (e.g. extent of dry rot or subsidence worse than anticipated).	Inadequate desk study or poor building investigation.
Shortcomings in discovery prior to commencement of contract (e.g. not establishing full extent of dry rot or subsidence).	Inadequate desk study or poor building investigation.
Structure and fabric in poorer condition than anticipated (e.g. opening up revealing more remedial works required than predicted).	Inadequate desk study or poor building investigation.
Structure and fabric lacking in performance requiring more extensive upgrade than originally anticipated (e.g. thermal insulation on flat roof thinner than expected).	Inadequate desk study or poor building investigation.
Obsolete or non-standard components difficult to match.	No local salvage source.
Fire outbreaks, particularly in old buildings.	Hot working processes on site such as welding, torching bituminous felt; arson or vandalism.
Reduction in fire resistance to building during refurbishment.	Fire protection removed – e.g. fire casing temporarily stripped from mild steel columns.
Water from rainwater leaks or leaking services causing damage to building fabric, finishings or contents.	Poor or inadequate protection against water.
Pest infestation more extensive/serious than expected.	Further pigeon colonisation in roof loft or rodent infestation.
Extent of deleterious materials (e.g. asbestos, woodwool slabs, etc) greater than expected.	Inadequate desk study or poor building investigation.
Retained portions of building requiring special consideration	Further repairs/restoration work required.
Unforeseen problems with mechanical services.	Premature breakdowns.
Need for protection of existing construction – additional protective measures to minimise collateral damage.	Building elements more vulnerable.
Party wall problems.	Uncertainty over ownership.
Working in occupied or unoccupied buildings – especially in hazardous areas.	Disturbance such as noise, dust; security problems. Confined spaces.
Need for partial demolition of existing construction.	Defect beyond that or worse than expected.

Table 2.7 Possible management and economic risks.

Example	Typical Cause
Changes in client requirements/finances.	Lack of communication.
Inaccurate or insufficient terms of reference.	Poor or vague brief from client.
Delays resulting in late decision making, late handing over of site, delayed programme.	Late deliveries, unforeseen problems/ defects.
Errors in design, contract documents, drawings.	Inadequate or lack of technical audit.
Failure to meet programme or timescale.	Many reasons in addition to those listed under row 3 above such as • not enough operatives on site; • poor programming or incompetent site supervision and control; • using inexperienced staff.
Estimating problems – inadequacies or inaccuracies resulting in escalating labour, plant and material costs.	Inadequate cost planning. Using inexperienced staff. Using out-of-date cost data Taxation changes.
Changes in project scope – occupancy, usage, size.	Clients changing their mind.
Site allowed to become untidy and unsafe.	Poor site supervision and control. Lack of labourers and other operatives on site to undertake regular/daily cleaning up operations.
Changes in site personnel.	New site manager inexperienced or unfamiliar with old buildings.
Contractual disputes and claims e.g. disturbance to neighbours.	Poor design team management.
Strikes or go-slows on site.	Poor industrial relations.
Difficulty in obtaining access or commencing work on programme.	Unco-operative occupiers.
Dangers for operatives working in potentially hazardous areas.	Confined spaces e.g. basements, roofspaces.
Bankruptcy/Liquidation of contractor/supplier/ design team members.	Adverse economic conditions Workflow problems.
Disruption of client's business being still undertaken in refurbished building.	Essential work required near to occupied parts of building.
Debris and dirt accumulation requiring frequent cleaning of site.	Neglect and untidy work practices.
Security breaches resulting in vandalism or theft of tools, materials, equipment, furniture, or possessions in building.	Inadequate site management and lack of security.
Occupation whilst work in progress and/or late request for partial handover.	Influences such as poor co-ordination of subcontractors, price changes permitted under certain contracts.

Table 2.8 Possible legal and commercial risks.

Example	Typical Cause
Non-compliance with the Building and Planning Regulations. Breaches of Listed Building requirements.	Cavalier or reckless work scheduling, particularly to old buildings.
Breaches of or changes to the regulations e.g. Construction/Building Regulations and Health and Safety Regulations.	Tightening up of statutory requirements e.g. DDA 2005, modifications to the CDM Regulations 2007. Failure to comply with technical requirements.
Health hazards to operatives, occupiers and general public.	Disregard of or non-compliance with safety requirements.
Lack of reporting and recording of accidents and safety incidents on site.	Non-compliance with RIDDOR regulations.
Freeholder and landlord requirements changing over time.	Unreasonable or difficult client or occupier.
Political change.	New political party in charge of national/ local government implementing fresh policies as regards construction and or planning. Although not retrospective, any new regulations may have an impact on a building's refurbishment.
Government legislation.	Inadequate desk study – not anticipating changes.Responding to EU and other Directives.
Interference/involvement of local/national pressure groups.	Local conservation bodies or national heritage agencies. Influence of bodies involved with disability requirements.
Legal agreements such as Rights of Way, Rights of Light, noise control requirements.	Neighbours enforcing their rights (per Party Wall etc. Act 1996).
Terrorist attack e.g. bomb blast or ramrod vehicle impact at/near an important public building.	Targeted building. Exacerbated by inadequate or slack security.

site. Several tragic accidents involving cranes in recent years have confirmed this, such as:

- On the afternoon of Saturday, 2 June 2007, part of a tower crane that was being extended fell on to the roof of the adjacent Croydon Park Hotel on Addiscombe Grove, in Croydon, London. The crane driver was seriously injured. The accident appeared to have been a repeat of the Canary Wharf incident listed below.
- In January 2006 a man died when a luffing jib crane collapsed on a David McLean site in central Liverpool, killing one site worker and trapping the operator in the cab.

- In September 2005 a tower crane in Battersea collapsed killing two people.
- Thornton (2005) reported that two workers were killed and a third injured on 11 February 2005 as a result of a crane collapse at a school building site in Worthing, Sussex.
- In 2000, three were killed by when a crane that was being extended collapsed in Canary Wharf, London.

These and many other crane incidents show that control of crane operations is not always properly managed. Clearly extending a crane is one of the major risk factors.

According to Thornton (2005):

> Where the project is a construction one covered by the Construction (Design and Management) Regulations (CDM) 1994 (now 2007) the project team – notably the planning supervisor (now known as co-ordinator) and principal contractor – will assist the client to ensure that operations involving cranes are safely controlled. However, there are a number of smaller projects that may not fall under CDM, which could cause the project manager controlling them to breach their duties of care if proper control is not taken.

(It should be noted that under the 2007 version of the CDM requirements all work falls under the Regulations. It is merely that small jobs are not notifiable.)

> The key piece of legislation covering cranes is the Lifting Operations and Lifting Equipment Regulations 1998. The main hazards that occur in use include:
> - Crushing – of a person by a swinging load or one that slips from its slings
> - Impact – on a person, structure or similar by the load or moving parts of the crane
> - Trapping – usually of the person slinging the load
> - People falling from the lifting equipment
> - Objects falling from the lifting equipment
> - Instability of the crane due to loose surface or inadequate ground strength
> - Collision of moving parts of two or more cranes
> - Crane collapse – usually during it being extended.
>
> Many of the issues will be covered by the company from whom the contractor hires the crane, and the skill of the competent operator of the crane – but you (the contractor) as the client must check and be able to prove you checked.
>
> However, the crane hirer cannot confirm things such as the state of the ground – is it strong enough to support the crane? Only you can advise this, probably in discussion with the hirer to determine the point loading of the crane.
>
> Similarly, you will know what activities are occurring on the vicinity that may interfere with the crane operations – again discussion with the hirer will help to determine the radius of operation of the crane that must be kept clear during the work.
>
> You as the client need to ensure that you carry out the right checks with the crane hirer, which must include:

- actually seeing the crane test certificates and records of thorough inspection and testing – one case involved a hirer falsifying these certificates and the client was held liable;
- ensuring the competence of the crane operator;
- ensuring that everyone is aware of who is actually in charge of the lifting operations – too many people signalling can cause confusion and incidents;
- what weather conditions must cause the crane operations to stop – and who will monitor these; and
- who will secure the clear area of safety of the radius of operation of the moving parts of the crane.

A good, properly chosen crane contractor will be able to guide you in suitable control of the crane operations, and reference to the British Standards series BS 7121 Code of Practice for safe use of cranes will also help.

It is important to note that oversailing of adjoining buildings requires the permission of the owner/s of the affected properties. This can take time and cost considerable sums of money.

2.8 Risk assessment

Again, according to the HSE's website:

Risk assessment is the process of identifying hazards, characterising the hazards, analysing the risks, evaluating the risks and determining the appropriate options for risk control.

In practice, it boils down to a careful examination of what, in your workplace, could cause harm to people, so that you can decide whether enough precautions have been taken or whether you need to do more to prevent harm.

The basic risk assessment procedure involves the following three main stages:

- Stage 1: Risk identification - sources and types of risk (Tables 2.5 to 2.8)
- Stage 2: Risk measurement - evaluation of risk: qualitative (high/medium/low); quantitative (1 to 5, see Table 2.15)
- Stage 3: Risk response - measures to eliminate, control, limit or transfer the risk (see below).

The HSE (2003a) prescribe five steps in the risk assessment process. Table 2.9 illustrates these five steps and indicates their implications.

Another factor to consider in risk assessment is the risk bearers. These are the stakeholder individuals who will be directly and indirectly affected by any hazard. This may be different from the liability for such risks. For example, contractors could be held responsible for failing to provide adequate Personal Protective Equipment (PPE) for their operatives (the immediate risk bearers).

Risk bearers fall into the following main groups: client, consultants (design team), contractors, operatives, occupiers and the public. Examples of the kinds of risk that these groups might bear are listed in Table 2.10.

Table 2.9 Risk assessment process (based on HSE 2003a).

Step	Requirements	Examples
1	Look for the hazards	Dangerous substances, working at elevated positions, etc.
2	Decide who might be harmed and how.	Operatives, staff, public, visitors.
3	Evaluate the risks and decide whether the existing precautions are adequate or whether more should be done	Use qualitative method initially. Do quantitative measurement if necessary. Increase precautions if necessary.
4	Record your findings	Compile a safety file and risk register (see below).
5	Review your assessment and revise it if necessary	Monitor and revise regularly.

Table 2.10 Typical risk bearers in refurbishment projects (Douglas 2006).

Risk Bearer	Examples of risks and hazards (some of which are shared between risk bearers)
Client	CDM Regulations seem to increase the client's responsibility regarding safety. Loss of use or interruption of use of building due to delays or overruns in the contract period. Increased repair costs or insurance premiums due to accidental or criminal damage. Lack of tenants or buyers for either the adapted building or units within the adapted building. Increased development and procurement costs.
Consultants	Negligence suit for alleged breach of professional duty – failure to inspect, design, specify, supervise or communicate properly. Injured whilst on site as a result of exposure to some of the same hazards as operatives. Loss or devalue of reputation as result of a bad publicity following a botched refurbishment scheme. Underestimation of fees or time required. Failure to obtain the necessary statutory approvals.
Contractor/s	Loss of production due to delays or overruns in the contract period. Disruption of work due to accidents or other safety scares on site. Failure to hire appropriately experienced/skilled operatives. Failure to obtain materials of adequate quality or at the correct time. Under-pricing work – likely leading to financial losses. Over-pricing work – possibly leading to failure to win contracts. Underestimating the time taken to complete work. Increased production costs.

Table 2.10 Typical risk bearers in refurbishment projects (Douglas 2006) *(continued).*

Risk Bearer	Examples of risks and hazards (some of which are shared between risk bearers)
Operatives (and personnel on site)	Injury or fatality in being exposed to a hazardous operation without proper PPE (see Table 2.3).Injury or fatality in being involved in an accident on site.
Occupiers and the public	Injury or fatality in being inadvertently exposed to a hazardous operation.Injury or fatality in being involved in an accident on site.

*Note:*The building itself may bear some risks, such as:Accidental damage to the structure/ fabric during adaptation work (e.g. inadvertent or careless removal of a loadbearing wall causing structural movement and cracking).Impact damage from vehicles or crane, or materials being lifted/dropped during crane operation.Flooding due to burst pipes (particularly in winter), leaking services or defective roof coverings/drainage.   Water ingress can cause extensive damage to valuable contents and could trigger fungal attack. Fire resulting from carelessly discarded cigarette, electrical fault or hot working process (e.g. improper or careless use of blow torches, which has caused major fire outbreaks in conservation works to some historic buildings).

Initially a qualitative risk analysis will usually be carried out because of its quickness and simplicity. This can be done using a brainstorming exercise. Figs. 2.4 and 2.5 illustrate typical risk grids. For example, in Fig. 2.4 activities or work involving a high probability and high impact merit a high priority in terms of risk control. Similarly, high risk activities or work have a high probability of occurrence and a high impact.

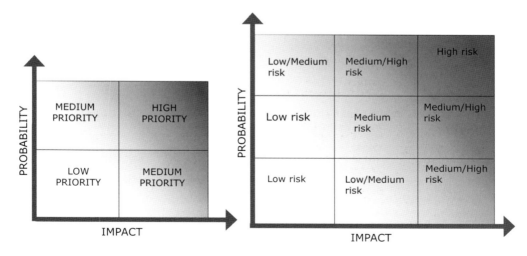

Figure 2.4 Simple risk grid.   **Figure 2.5** Risk matrix.

A useful way of quantifying the technical risks identified would be to use a risk schedule as shown in Table 2.11. It is derived from the major refurbishment and repair of a Grade A (I) Listed public building in the heart of Edinburgh, UK (see also Chapter 7). The project is described in more detail in the case study outlined in Appendix A.

The evaluation of likelihood of occurrence (i.e. probability) in Table 2.8 was done using a simple quantitative method. The probability risk factor indicated can be derived from the probability gradings shown in Tables 2.12 and 2.13.

Some classification of the impact of risks can be further quantitatively assessed. Table 2.14 shows how this could be done.

Table 2.11 Risk schedule estimating maximum risk allowance for a large public building refurbishment project (Douglas 2006).

Refurbishment Option: North/South Phasing of Building			Average Risk	
Risk Element (1)	Type of Risk (2)	Base Value of Risk Element	Probability Factor (4)	Assessment (5)
Rot repairs	V	(3)	0.95	23,750
Additional masonry repairs	V	25,000	0.50	15,000
Additional roof repairs	V	30,000	0.90	22,500
Asbestos abatement	V	25,000	0.50	25,000
Structural alterations	V	50,000	0.50	150,000
Fire protection	V	300,000	0.75	15,000
Fire officer requirements	V	20,000	0.40	40,000
Safety officer requirements	V	100,000	0.50	12,500
Funding body's requirements	V	25,000	0.50	5,000
Restrictive working methods	V	10,000	0.75	30,000
Penalty clause	F	40,000	0.25	2,500
		Average risk allowance (6)		£341,250
				Say £342,000

Notes:
(1) Determined by the nature of the refurbishment project and type and condition of building.
(2) F = Fixed (e.g. penalty clause), V = Variable (e.g. dry rot repairs).
(3) Based on approximate cost of remedial works on similar size buildings or previous jobs of a similar scale.
(4) A high probability is 95%; this is represented as a probability factor of 0.95.
(5) This is simply (3) x (4).
(6) This figure can form the basis of the minimum contingency sum for the contract.

Table 2.12 Simple probability ratings (based on Edwards 1995).

Assessed likelihood	Equivalent probability	Approx. qualitative rating
No chance of occurring	0%	Low
Unlikely to occur	5–45%	
As likely as not	45–55%	Medium
Likely	55–95%	
Almost certain	95–99%	High
Certain to occur	100%	

Table 2.13 Basic qualitative/quantitative probability ratings (based on Edwards 1995).

Assessed likelihood	Equivalent probability	Approx. qualitative rating
Loss is not possible	0	Low
Very remote possibility	0.1	
Remote possibility	0.2	
Slight chance of occurrence	0.3	
Slightly less than equal chance	0.4	
Equal chance of occurring	0.5	Medium
Fairly possible	0.6	
More than likely to occur	0.7	
Predictable	0.8	High
Almost certain	0.9	
Loss is certain	1.0	

Table 2.14 Severity grading for risk analysis (Bowles and Kelly 2005).

Grading	Assessment of impact	Estimated cost*
1	Minimal impact, nuisance only	£
2	Medium loss	£
3	Manageable loss	£
4	In range of largest previous loss	£
5	Serious loss	£
6	Catastrophic	£

*Depends on the size and duration of contract, type and location of building, etc.

2.9 Risk profile

Assessing the risk characteristics of a refurbishment project can be done by using a risk profile. This is basically a checklist of factors that are used to assess the overall 'riskiness' of such a project. Understanding all the elements of the risk anatomy of a project helps to heighten awareness of, and sensitise the design team to, risk.

A typical risk profile for a refurbishment project is illustrated in Table 2.15. It shows a range of factors against which risks can be assessed.

The factors listed in Table 2.15 can be grouped according to whether they are strategic, tactical or operational. Table 2.16 shows such groupings.

Table 2.15 Risk profile of a refurbishment project (based on Baccarini and Archer 2001).

Factor	High		Risk Rating		Low
	5	4	3	2	1
Uniqueness of project	Prototype incorporating new techniques	Unusual project	Conventional project	Modifications to existing design	One of a series of repetitions
Complexity of deliverable	Outcome based contract (eg, PFI)	Coordination of services (eg, FM)	Design and build	Supply and installation	Supply only
Financing	Private sector funding or joint venture	Capital works not yet approved or requested	Capital works in forward estimates	Capital works already allocated	Recurrent funds in current year
Adequacy of funds	Very likely to be inadequate	Likely to be inadequate	Tight budget, achievable with control	Adequate with some contingency	Adequate with generous contingency
Building location	Remote, inaccessible	Remote but accessible	Regional but distant	Regional	Metropolitan
Building surroundings	Activities in occupied areas	Staging within occupied areas	Additions to occupied areas	Near some occupied areas	Clear of occupied areas
Deleterious materials	Working with hazardous materials	Possibly involves hazd. materials	Hazd. materials exist but not part of works	Unlikely to encounter hazd. materials	No known hazardous materials
Definition of project	No project information available	Brief project description	Generic project brief available	Feasibility study completed	Detailed project brief available
Building availability	Building not identified	Several buildings identified	Building identified but not yet purchased	Existing building purchased	Existing building already owned
Project justification	No need has been justified	Justification is questionable	Need justified but may change through project	Need justified based on historical information	Need fully justified through recognised process

Table 2.15 Risk profile of a refurbishment project (based on Baccarini and Archer 2001) *(continued)*.

Factor	High ◄──────────	Risk Rating	──────────► Low		
	5	4	3	2	1
Project approvals	Unidentified approvals required	Potential approval delays have been identified	Required approvals are known and documented	Few approvals required or most obtained	No approvals required or already obtained
Client's experience	Inexperienced multiple clients	Mixed experience amongst clients	Inexperienced single client	Experienced multiple clients	Experienced single client
Client's relationships	Multiple reluctant clients or relationship not established	Mixed relationship with clients	Reluctant client or relationship not yet established	Good working relationship (multiple clients)	Good working relationship with single client
Assessment of contractors	Unknown contractors	Limited number of unknown contractors	Limited number of competent contractors	Adequate number of competent contr.	Abundance of competent contractors
Procurement method	No tendering and involving sponsorship	Negotiated tender	Tendered outside agency	Public open tender	Selected tender
Consultant selection	Selection without approved processes	Design competition	Selection from limited list of consultants	Period panel consultant	Consultant selected using approved process
Stakeholder interest	High level of political, community or media sensitivity	High profile client or project	Stakeholder groups involved	Project may attract stakeholder or media interest	Project unlikely to attract stakeholder or media interest
Other, project-specific, factors ...					

2.10 Risk response

Naturally, it may not be possible or appropriate to instigate responses to all identified risks. The refurbishment design team should therefore predominantly concentrate on the higher priority areas. The primary measures to cancel or control negative risk are summarized in Table 2.17.

Table 2.16 Grouping of risk factors (based on Bowles and Kelly 2005).

Context	Examples	Level	Parties Involved
Project related	Funds/financing Definition of a project Project justification Stakeholder interest Building availability	Strategic (1)	Client Consultants Approving authorities
	Property location Project surroundings Hazardous materials Availability of contractors/suppliers Uniqueness of product	Tactical (2)	Consultants Contractors Sub-Contractors Suppliers
	Day-to-day on-site activities Safety checks and inspections Site safety meetings	Operational (3)	Contractors
Project Management related	Client's experience Client relationships Consultants selection Safety Co-ordinator appointment	Various	Various

Notes
(1) Strategic phase covers the initial stages of a project concerned with defining its scope and developing the brief. (Long term issues.)
(2) Tactical phase covers the main stages of a project concerned with its design and delivery. (Medium term issues.)
(3) Operational phase covers the implementation stages of a project concerned with day-to-day issues during the construction period. (Short term issues.)

Table 2.17 Range of risk responses.

Response	Examples
Elimination	Change to a less hazardous operation (e.g. using cold-applied roof membrane instead of hot-applied covering for flat roof refurbishment work to an existing high risk building such as a school). Find an alternative to undertaking the proposed hazardous work inside the building being refurbished. The contractor will usually be asked for a solution to the problem.
Reducing probability of occurrence	Safety management regime (HSE 2003b): A coherent and enforced safety policy; regular safety briefings. Strict smoking ban on site. Hazardous chemicals and other dangerous materials stored in a safe, secure location. Audit, review and feedback: Site meetings to address safety issues. Risk monitoring: Regular checks by all staff, especially the safety co-ordinator. 24 hour site security. Other site precautions (see below)
Reducing impact of occurrence	Standby generator. Emergency evacuation plan. Fire drills. Installing a sprinkler system in the building or fitting appropriate fire extinguishers (FE) within the building at key positions (e.g. CO_2 FE next to electrical equipment and machinery). Other site precautions (see below).
Transfer of risk	Transfer risk to another party (eg, to the relevant sub-contractor). Insurance: General property insurance (GPI) for monetary loss not criminality; and professional indemnity insurance (PII)

All individuals on well managed sites (i.e. members of staff, operatives and visitors) are required to wear the following three pieces of personal protective equipment (PPE):

- British Standard approved hard hat
- high visibility vest
- safety boots/shoes

Repeated failure to do could result in the person concerned being expelled and banned from the site for a restricted period or until the contract is completed. This normally occurs only after a third breach of such a requirement by an individual on site.

On large refurbishment or new-build schemes there is now a requirement for a permit to be gained by examination before a person can visit the site. Other typical safety precautions and facilities on a building refurbishment project (Douglas 2006) can be classified into physical and procedural measures. These are summarized in Tables 2.18 and 2.19 respectively. Table 2.20 provides a sample checklist that can be used in a hazard assessment exercise relating to refurbishment work. (See also HSE 2003(b).)

Table 2.18 Physical safety precautions to reduce site risks (based on Douglas 2006).

Precaution	Example
Adequate PPE over and above the three items referred to above must be worn by all operatives when undertaking hazardous operations such as cutting, drilling, scraping, working with toxic substances etc.	Goggles or safety glasses/visor, face masks, safety gloves, overalls, etc.
Properly rated and earthed electrical equipment, with circuit breakers or emergency cut-off switches nearby. A qualified electrician must check these at least once a year.	Only certified and regularly tested equipment should be used.
Dust control measures.	If dust creation is unavoidable, ideally only high efficiency particulate arrestor (HEPA) filters should be used in face masks and vacuum cleaning equipment where protection against or control of dust and microscopic contaminants such as bacteria, mould and other airborne contaminants is required. A HEPA filter is a filter that removes 99.97% of particulates 0.3 microns or larger in size (Thompson 2005). Dampen down surfaces.
Bright/luminous padding should be fitted around scaffolding poles or other protruding points at dangerous or vulnerable installations	Resilient polyethylene foam preformed sections covered with red-white striped bands with conspicuous (red and white striped) warning stickers around scaffold tubes. These are especially needed at and below 2.5m (ie, within head height).

Table 2.18 Physical safety precautions to reduce site risks (based on Douglas 2006) *(continued).*

Precaution	Example
Adequate harnesses and crawlboards to protect operatives working on pitched or flat roofs.	Safety anchor fixings and movable working platforms particularly on buildings with fragile coverings.
Adequate secured and appropriate fixed or portable scaffolding on proper support.	Anchored at every storey and on stable baseplate.
Adequate and robust barriers on scaffolding and other platforms above 1m.	Stable balustrades; kicker boards.
Temporary ramps at changes in floor level – to reduce the number or presence of steps.	Stable and robust platforms with hardwood or metal ribs securely fixed to ramp surface to minimise slips and trips.
Good quality ladders secured and properly positioned.	Adequately tied and at the correct incline.
Waste chutes constructed to transfer debris safely down scaffolding to a skip.	Using articulated plastic pipes c. 400 mm in diameter or plywood ducts the same size.

Table 2.19 Procedural safety precautions to reduce site risks.

Precaution	Example
Good housekeeping and recycling are essential for a safe and environmentally friendly site.	The building being adapted must be kept clean and tidy on a daily basis. Rubbish and construction waste should be removed (in sealed bags where possible) and deposited in a suitable skip. Any rubbish containers such as skips need to be emptied regularly to an approved dump. Waste should be separated to maximize recycling.
Dust management measures (see Douglas 2006).	Restrict or regulate the timing and extent of any cutting, drilling and other dust-generating activities. Dampen down dust-laden areas with fine water spray,
Adequate evacuation procedures and means of escape clearly identified.	Regular fire drills and safety training.Dead ends when installing flammable materials, such as adhesives for floor tiles, should be avoided.Alarms should be tested regularly.
Strict controls and close supervision of hot working and other activities involving flammable processes.	Hot working permit required. Appropriate portable fire extinguishers should be readily available nearby such work.
Timing hazardous operations outside core business hours to minimise disruption and nuisance.	Asbestos removal, hot or noisy and dusty work done at night or weekends if necessary.
Flammable materials and low-pressure gas cylinders should be kept in secure, adequately ventilated enclosures or stores.	Separate site hut or designated room within the building being refurbished.

Table 2.19 Procedural safety precautions to reduce site risks *(continued)*.

Precaution	Example
Create separate access routes for occupiers and the public so that they do not intermingle with site operatives.	Large notices and direction signs should be placed at doorways to indicate permitted users and directions.
All operatives and staff on site are adequately briefed and regularly updated as to safety precautions and emergency procedures.	To facilitate this large safety notices should be placed at conspicuous points around the building.
Temporary support mechanisms before and during partial demolitions works.	Adequate anchoring/restraint/support for propping and other shoring.

Table 2.20 Extract of sample checklist based on a medium-size refurbishment project.

DESCRIPTION OF HAZARD	Risk Factor Ratings						ACTION
	EX	PR	SEV	PRO	RES	ACC	
(1) Security of site against unauthorized entry	No	Yes	M	M	M	Yes	Contractor to ensure that site is secured at end of each working day and that all appropriate signage is provided.
(2) Potential slipping/falling	No	Yes	M	M	M	Yes	Contractor to ensure that due care is taken at all times and that full scaffolding with all necessary edge protection is provided.
(3) Existing hazardous materials on site	Yes	Yes	L	L	L	Yes	Contractor to ascertain all hazardous materials prior to carrying out work and take all necessary precautions.
(4) Dust and debris arising from cutting of materials (e.g. MDF board, timber, tiles, etc.)	No	Yes	M	M	M	Yes	Personal protective equipment to be issued to and worn by all site operatives as necessary.
(5) Working with lead	Yes	Yes	M	M	M	Yes	Contractor to take all necessary health and safety precautions in accordance with approved Regulations and HSE Information Sheet No. 4 when working with lead.

Table 2.20 Extract of sample checklist based on a medium-size refurbishment project *(continued)*.

DESCRIPTION OF HAZARD	Risk Factor Ratings						ACTION
	EX	PR	SEV	PRO	RES	ACC	
(6) Working with live electrics	Yes	No	H	M	H	Yes	Contractor is to exercise care at all times when working with electrics. Permit to work system to be adopted where necessary.
(7) Paint fumes	No	Yes	M	M	M	Yes	All protective equipment/ clothing to be worn when working with solvent-based paints.
(8) Working with blow torch on roof repairs	No	Yes	H	M	H	Yes	All protective equipment/ clothing to be worn. Obtain hot working permit prior to the commencement of this work.

KEY & NOTES

EX = Existing Hazard
PR = Hazard arising from proposed construction
SEV = Severity of risk
PRO = Probability of risk (High/Medium/Low)
RES = Result (High/Medium/Low)
ACC = Accept (Yes/No) risk?
Severity x Probability = Result
H x H/H x M/M x H = H
H x L/L x H/M x M/M x L/L x M = M
L x L = L

ACTION

What has been done
What should be done
Legislation/Ref to Plan/Ref to File

H = High   M = Medium   L = Low

2.11 Risk register

Any well managed refurbishment project should have a risk register (Smith *et al.*, 2006). This is a record of all the known risk activities on site. A copy should be kept by the site agent or project manager as well as the safety coordinator. Fig. 2.6 shows an extract from a typical risk register.

2.12 Summary

As was pointed out in Chapter 1, refurbishment and maintenance work account for nearly half of the UK construction industry's output (Douglas 2006). This sector of the

Figure 2.6 Typical risk register (based on Bowles and Kelly 2005).

Item No.	Source	Owner	Consequence	Probability of occurrence	Response	Effect		
						Eliminate	Transfer	Reduce
1.0	Laying asphalt tanking in basement	Contractor	Fire spread, melting adjacent materials, noxious fumes	0.95	Hot working permit; restricted working hours	No	No	Yes

construction industry has more than its fair share of risks, particularly those relating to health and safety.

Risk assessment and risk management, therefore, are essential requirements for any safe and efficient construction project (Holt 2005), including refurbishment. Their recognition and full implementation will go a long way to minimise if not prevent the accidents, damage and injuries that may occur in a building undergoing such work.

This chapter provides an outline of the basic methodology for checks that can be used to achieve that goal. The construction industry and society as a whole deserve nothing less.

Bibliography and further reference

Baccarini, D. and Archer, R. 2001. The risk ranking of projects: a methodology, *International Journal of Project Management*, 19(3), Elsevier, Oxford.

BCIS. 2005. *The Economic Significance of Maintenance (SR 338)*, Building Cost Information Service, London.

BCIS. 2005. *BCIS Guide to House Rebuilding Costs*, Building Cost Information Service, London.

Bowles, G. and Kelly, J.R. 2005. *Value and Risk Management.* D19CV9 Course Notes for MSc Construction Management programme, Heriot-Watt University, Edinburgh.

Blockley, D. (Ed.) 1992. *Engineering Safety*, McGraw-Hill, London.

BSI. 2002. *Risk Management vocabulary – Guidelines for use in Standards (Guide 73)*, BSI, London.

CIRIA. 2002. *A simple guide to controlling risk,* CIRIA, London.

Construction Planning and Procurement Panel. 2000. *The Management of Risk*, RICS Information Paper, RICS Books, London.

Dallas, M.F. 2006. *Value and Risk Management: a guide to good practice*, Blackwell Publishing, Oxford.

Doran, D. 2009. Construction Plant. In: Doran, D. (Ed.) *Site Engineers Manual*, 2nd edn., Whittles Publishing, Scotland.

Douglas, J. 2006. *Building Adaptation*, 2nd edn., Butterworth-Heinemann, Oxford.

Douglas, J. and Ransom, W. 2007. *Understanding Building Failures*, 3rd edn., Taylor and Francis, Oxford.

Edwards, L. 1995. *Practical risk management in the construction industry,* Thomas Telford, London.

Egbu, C.O. 1995. Perceived degree of difficulty of management tasks in construction refurbishment work, in *Building Research and Information*, 23(6), E and FN Spon, London.

Euroroof Ltd. 1985. *Re-Roofing: a Guide to flat roof maintenance and refurbishment*, Euroroof Ltd., Northwich.

Ferguson, I. 1995. *Dust and noise in the construction process*, HSE Contract Research Report 73/1995, HMSO, London.

Flanagan, R. and Norman, G. 1993. *Risk Management in Construction*, Blackwell Science, Oxford.

Hambly E.C. and Hambly, E.A. 1994. Risk Evaluation and Realism, *Proceedings of the ICE, Civil Engineering*, 102(2), Thomas Telford, London.

Harris P. 2009. Anthrax in conservation work. *The Structural Engineer*, 87(11), IStructE, London.

Hillson D. 2002. Extending the Risk Process to Manage Opportunities, *International Journal of Project Management*, 20(3), Elsevier, Oxford.

Historic Scotland. 2001. *Fire Risk Management in Heritage Buildings,* Technical Advice Note 22, Historic Scotland, Edinburgh.

HSE. 2001. *Reducing risks, protecting people: HSE's decision making process*, HSE, London.

HSE. 2003(a). *Five steps to risk assessment,* HSE. Leaflet; available online at http://www.hse.gov.uk/pubns/indg163.pdf. Accessed on 01 June 2009.

HSE. 2003(b). *Managing health and safety: five steps to success*, HSE, http://www.hse.gov.uk/pubns/indg275.pdf, Accessed on 01 June 2009.

Lead Sheet Association. 1990. *Condensation – a problem even for lead*, Lead Work Technical Notes 3, Lead Sheet Association, London.

McGuinness, P. 1995. *Risk assessment: a line manager's guide,* The Industrial Society, London.

NMAB. 1982. *Conservation of Historic Stone Buildings and Monuments*, Report of the Committee on Conservation of Historic Stone Buildings and Monuments, National Academy Press, Washington DC.

Noy, E. and Douglas, J. 2005. *Building Surveys and Reports,* 3rd edn., Blackwell Publishing, Oxford.

OGC. 2002. *Management of Risk: guidance for practitioners book*, Office of Government Commerce, Norwich.

OGC. 2007. *Procurement Guide 03: project procurement lifecycle the integrated process*, Office of Government Commerce, http://www.ogc.gov.uk/documents/CP0063AEGuide3.pdf. Accessed 2 February 2009.

OGC 2007. *Procurement Guide 04: risk and value management*, Office of Government Commerce, http://www.ogc.gov.uk/documents/CP0064AEGuide4.pdf. Accessed 2 February 2009.

Perry, P. 1999. *Risk assessment questions and answers*, Thomas Telford, London.

Raftery, J. 1994. *Risk analysis in project management*, E and FN Spon, London.

RICS. 2000. *The management of risk*, An Information Paper, RICS Business Services, London.

Royal Society. 1983. *Risk Assessment,* A Study Group Report, The Royal Society, London.

Royal Society, 1992, *risk: analysis, perception and management*, A Study Group Report, The Royal Society, London.

St. John Holt, A. 2005. *Principles of construction safety*, Blackwell Publishing, Oxford.

Sawczuk, B. 1996. *Risk avoidance for the building team*, E and FN Spon, London.

Smith N.J. Jobling P. and Merna, T. 2006. *Managing risk in construction projects*, 2nd edn., Blackwell Publishing, Oxford.

Teo, H.P. 1991. *Risk perception of contractors in competitive bidding for refurbishment work*, RICS Books, London.

Thompson, K. 2005. *Get Mould Solutions*, http://www.getmoldsolutions.com/hepa_vacuums_exposed.html. Accessed 1 March 2009.

Thornton, E. 2005. *Crane safety for occupiers of premises*, http://www.workplacelaw.net. Accessed 1 March 2009.

Ward, S. and Chapman, C. 2003. Transforming project risk management into project uncertainty management, *International Journal of Project Management*, 21(2), Elsevier, London.

Weatherhead, M., Owen, K. and Hall, C. 2003, *Integrating value and risk in construction*, CIRIA, London.

3 Discovery: including sources of information

3.1 General

Before embarking on a scheme it is imperative that a thorough forensic-style investigation of the existing building or structure is carried out. Approaches may need to be made to:

- building or structure owners;
- former owners;
- original designers (architects, engineering consultants) and designers of any subsequent alterations;
- contractors and specialist subcontractors; and
- statutory organisations.

In dealing with material gathered from discovery searches, professionals have found it convenient to divide this into 'primary material' (original construction drawings, specifications etc) and 'secondary material' (case studies and other similar matter).

(In recent years many architectural and engineering practices have been taken over by or merged with larger organisations. To assist in tracing the location of engineering firms The Institution of Structural Engineers has made available a tracker. This tracker can be viewed on the Institution's web page.)

- Site investigation material. For larger projects it might be worthwhile contacting the Geological Society (GS) who maintain a database of site information material from earlier projects.
- Age of original construction and of subsequent alterations.
- Types of construction (see Chapter 5).
- History, i.e. construction dates of any major alterations since initial construction.
- Condition of existing structure, e.g. building and structural surveys.
- Defects, e.g. waterproofing, structural etc.
- Dangerous substances, e.g. asbestos.
- State of mechanical services.
- Status, e.g. listed building (see Chapter 6).
- Party structures, e.g. etc.
- Original design drawings, specifications etc.

- Health and Safety files for jobs after the introduction of the CDM Regulations in 1994 (revised 2007). CDM regulations also required the production of as-built drawings. Experience has shown, however, that these may not have been prepared and if available may be of doubtful accuracy. If doubts concerning the accuracy of these exist further investigation is essential.
- Data from original fire precaution assessments.
- Local Authority information. In central London the District Surveyor still maintains an archive of original documents. In outlying districts, though, the picture is very patchy. However, it is worthwhile approaching the local building control officer who may have a file of documents relating to a particular development. This type of documentation may have been transferred to microfiche/microfilm or similar media.
- Site listing, including archaeological interest.
- Site restrictions, e.g. is it a reclaimed brown field site?
- Restrictions on the use of explosives for demolition etc.
- Working time agreements.
- Underground issues, e.g. Fleet and/or Tyburn underground rivers; Post Office Tunnel, Pneumatic Post Office Railway Tunnel, London Underground Ltd. (LUL), British Rail/Railtrack etc.
- Ancient lights issues, e.g. common service supply line agreements, etc. under the Party Wall etc. Act.
- Archival material.

3.2 The role of testing and monitoring in the discovery process

Properly designed and executed tests can play a big part in the understanding of existing structures.

3.2.1 General

Testing may be considered in two classes, non-destructive (NDT) and intrusive. NDT usually involves visual inspection assisted by such techniques as subsurface radar, infra-red photography and ultrasonic methods. These techniques have been used, for example, to check the adequacy of fire compartmentation walls in Buckingham Palace and at Heathrow Airport to assess ground stability and location of voiding beneath a concrete slab after the catastrophic collapse of a tunnel near a terminal building. A fuller description of these techniques can be found in *Non-destructive investigation of standing structures* (Historical Scotland, 2001).

Non-destructive testing can be conveniently grouped into three generic categories:

- Electro-magnetic methods (impulse radar, thermography, metal detection, free electro-magnetic radiation. Thermal imaging is particularly useful in checking for the presence and distribution of wall ties in cavity walling, thermal resistance and the presence of old rubble fill in the fabric of historic buildings.
- Nuclear (or X-ray) methods (radiography). These technique can be used, for example, on steel to ascertain the quality of welds, the presence of cracks, laminations, porosity and inclusions.
- Mechanical methods, such as ultrasonic pulse velocity (UPV), impact-echo. Impulse-echo, also known as Impulse Radar or Ground Penetrating Radar (GPR), can be used to map the arrangement of elements within a structure. UPV was particularly useful in checking the adequacy of prestressed concrete units following the problems with high alumina cement in the 1970s.

These non-intrusive investigatory techniques are increasingly being used in the UK, Mainland Europe, USA, Australia, India and elsewhere.

It is sometimes appropriate to load test an existing building or structure to test for structural performance. This may be necessary when original construction drawings are

Impulse radar

Impulse radar testing to locate bars. A radio wave produces an echo sounding pulse that can be utilised in the internal assessment of a wide variety of construction materials. It is particularly appropriate when applied to concrete.

Here amongst electrical switchgear where covermeters could not work, data on depth, size, location and condition are collected easily. The records are permanent and data can be mapped to build complete as-built detailing. (Courtesy GB Geotechnics Ltd., Cambridge.)

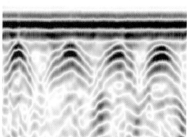

missing or inadequate or when a structure has been damaged due to fire, impact or explosion. In such tests loads will usually be applied incrementally and deflections monitored (see below for additional applications of monitoring as an adjunct to testing). It is essential not to overload or collapse a structure thus inhibiting its subsequent use without extensive repair.

In carrying out such tests the Health and Safety of operatives is paramount.

3.2.2 Specific material tests that may be relevant to refurbishment and repair work

3.2.2.1 Concrete

- Rebound hammer tests: provide a comparative assessment of concrete strength by testing its surface hardness.
- UPV tests: assesses the quality and uniformity of concrete.
- Core testing: used on 100–150 mm diameter cores to measure the strength and density of concrete.
- Internal fracture tests on concrete: provide a measure of the compressive strength by inducing internal fracture within the material.
- Windsor probe test on concrete: provides an empirical estimate of the compressive strength by firing a steel pin into the material.
- Break-off test: provides a measure of the tensile strength of concrete by applying a transverse force to the top of a core.
- Phenolphthalein test to the broken surface of concrete: determines the reduced alkalinity caused by the penetration of acidic atmospheric gases such as carbon dioxide. It should be noted that this test is not applicable to High Alumina Cement (HAC) concrete.
- Microscopy studies: determine the constituents, voids, mineralogical formation and other structural features by examination of thin sections of concrete or mineral materials.
- Free lime content – depth of carbonation (*qv*): determines the depth to which the concrete has been affected from the exposed surfaces by laboratory analysis.
- Cement content and cement/aggregate ratio of concrete: a sample analysis from a chemical laboratory.
- Types of cement in concrete: may be apparent from visual examination (for

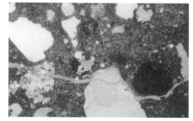

Microscopy

A thin section of concrete used in petrological analysis. This has been prepared from a concrete core sample impregnated under vacuum with a coloured resin. It shows a crack passing around and through different aggregate particles.

example the dark grey or brown colour of HAC) but if doubt remains thin section analysis should clarify the result.

- Water/cement ratio in concrete: may be approximately determined from samples in the laboratory by saturation techniques. More accurate results may be deduced from thin section analysis.
- Initial surface absorption test (ISAT): needs to be carried out on oven dry concrete samples to measure surface absorption in order to assess durability.
- Water and gas permeability tests on concrete: assesses the permeability of the surface zone – usually on samples in the laboratory. Included in this suite of tests are Figg and CLAM water and air permeability tests.
- Absorption tests on small (75-mm diameter) cores: determine absorption limits at different ages, sometimes required for precast concrete products
- Chloride tests on drilled concrete and mortar samples: commercially available kits can be used to determine chloride levels that may be present, for example, due to dosing in the construction phase to accelerate early strength gain.
- Tests for admixtures and contaminants such as chlorides, sulfates and other chemical materials in concrete: determined by laboratory techniques such as X-ray fluorescent spectroscopy, infrared absorption and scanning microscopy.
- Moisture measurements for concrete, masonry products and timber: the most direct method is to weigh a sample (W1), dry it in an oven then re-weigh (W2). The moisture content is then $W1 - W2 / W2 \times 100\%$.
- Abrasion resistance testing (usually for concrete): an accelerated wear apparatus consisting of a rotating loaded plate supported by three case hardened steel wheels which wear a groove in the concrete surface can be used.
- Testing for air entrainment for concrete: may be established by microscopy point counting methods on prepared samples impregnated with a suitable dye.
- Covermeter tests to establish the depth of cover to reinforcement in concrete: also helpful in determining the orientation and distribution of reinforcement.
- Electrical potential: measure the electrical potential of embedded reinforcing steel relative to a reference half cell placed on the concrete surface.

3.2.2.2 Timber

- Visual examination of timber: can be used to identify timber species, grade and quality, insect and fungal attack and condition of joints.
- Identification of insect attack: the commonest pests infesting constructional timber are the common furniture beetle and the longhorn beetle. (The latter is prevalent only in parts of the south-east of England.)

- Identification of dry and/or wet rot: the early signs of fungal attack are not easy to detect without laboratory equipment. The early stages of decay are usually easier to identify than the fungus when timber has been subjected to prolonged periods of moisture contents in excess of 20% and inadequate ventilation. Early signs of dry rot are cellular-type cracks along the grain followed by cracks at right angles to the grain. It should be noted that it can also progress through other materials such as brickwork. Wet rot continues to develop only on timber that is wet, whereas dry rot, having established itself on wet timber, will spread to otherwise sound dry timber. Accurate diagnosis is essential as treatment for the two conditions differs.
- Moisture control of timber: the use of portable battery-operated moisture meter on solid untreated timber is usually sufficiently accurate to determine moisture content if manufacturer's correction factors are correctly applied. If moisture content has risen to 15–16% it might be prudent to check temperature and humidity to see if heating and ventilation of the area is adequate.
- Mechanical properties of timber: in the absence of markings, stress grading of timber can be assessed visually or mechanically. The former requires the presence of a skilled and experienced practitioner who can measure defects and other features, and from those observations deduce the stress grade of the material. Mechanical grading requires standard samples to be processed through the appropriate equipment.
- Identification of glues/adhesives: chemical analysis is usually required, however, rules of thumb can provide a rough guide. For example if the glue is dark brown it is likely that the glue is a weather- and boil-proof glue such as resorcinol. If it is white then it could be either a moisture-resistant glue such as urea or – for interiors – casein.
- Identification of preservative treatments: some treatments can be recognised by colour and possibly odour. For example creosote has a distinctive smell and brown colour. Green-coloured timber may suggest treatment with a copper containing formulation such as copper carboxylates. A pale green-brown colour may indicate the presence of copper/arsenic/chromium formulations. Most reputable manufacturers should be able to recognise the presence of their particular formulation. Laboratory tests on a small sample (which includes sapwood) should reveal the true identity of a preservative.

3.2.2.3 Masonry

- Crushing of masonry cores, units or sawn-out samples including mortar joints: dry coring techniques can be used in brick and concrete block masonry. Samples for testing may also be obtained by removing bricks or blocks from the masonry or by sawn-out masonry samples. Strength may be determined by crushing cores or other types of samples in the laboratory.

- Helix pull-out test (still under development): gives an indication of the compressive strength of mortar or lightweight aerated blocks.
- Split cylinder tests: measures lateral tensile failure which is influenced by the mortar undergoing greater transverse deformation than the masonry units
- Flat jack test: measures the *in-situ* compressive stress in masonry using a special flat jack inserted into a horizontal slot which is cut into a bed joint.
- Endoprobe and boroscope observations: can be used to inspect the integrity of cavity wall ties or other elements within cavities, for example, beneath suspended timber floors.

3.2.2.4 Metals

- Visual identification of cast and wrought iron: cast iron may be recognised by the gritty surface texture of the material. It often occurs in thick or coarse sections with tension flanges larger than compression flanges. Wrought iron is not easily distinguishable from low-carbon steel and may require testing of samples to determine tensile strengths to confirm the material.
- Chemical analysis of metals: laboratory analysis can be used to determine origin and physical properties.
- Metallography: examination on small sample (e.g. 10 mm × 10 mm) to determine the internal structure of a metal, which may, for example, be required following a weld failure.
- Dye penetrants: techniques used to provide information on the surface condition of steel (or on the welds themselves) to show up imperfections that might, for example, affect the quality of welding.
- Ultrasonics – steel and other metals: can be used to indicate the presence of laminations and lamellar tearing although the coarse grain of some cast iron and the laminated structure of wrought iron may limit the application of ultrasound in these materials
- Radiographic techniques for metals: used to determine the integrity of steel and other metals (for example to ascertain the presence and degree of cracks, laminations, porosity and inclusions, both in parent materials and welds).
- Hardness tests: hardness of a material is determined from the size of an indentation made on its surface. The hardness numbers can be determined by a variety of laboratory tests such as the Brinell, Vickers or Rockwell tests.
- Tensile tests: measures the strength of a material by rupturing a standard specimen and can also be used to obtain the ductility of the metal. Test pieces (usually circular) generally vary in length between 150 mm and 250 mm.
- Wedge penetration test for cast iron: requires a disc (25–50 mm in diameter) which is placed between a hardened wedge and an anvil. The

force on the wedge at which the disc splits divided by the area of the split (the splitting strength) can be calibrated and used as a measure of quality control. The test has been used as part of the assessment of the tensile strength of the cast iron in a bridge. Further information may be obtained from the British Cast Iron Research Association (BCIRA).

- Split cylinder test for cast iron: a compression test on a small machined cylinder of cast iron placed horizontally between the platens of a testing machine and compressed until it splits along a vertical diametric plane.
- Impact tests: measure the energy required to fracture a standard notched specimen with a blow from a pendulum. There are two main types of test: the Charpy (beam) and the Izod (cantilever) test, of which the former is more versatile as it enables results to be obtained over a range of temperatures. These tests enable the notch ductility of the metals to be measured. Samples should be of 10 mm × 10 mm cross-section and 55–126 mm long, depending on the type of test and the required results.
- Visual examination for weld defects: surface appearance of welds may give an indication of the quality of the welding and the presence of defects such as cracking. The surface must be clean and in this connection the use of dye penetrants may be helpful.
- Magnetic-particle crack detection: mainly used during the fabrication and erection of steelwork and can only detect surface and near surface defects in metallic materials.
- Chemical tests of bronzes: allow analyses to be made on drillings to determine the composition of the metal. Special tests on small solid samples can indicate susceptibility to stress-corrosion cracking.
- Condition of steel cables (e.g. structural cables used in bridges): inspection is possible by visual examination bearing in mind the following:
  - Damage is most likely to be present at the ends (terminations); the outer wires will generally fail before the inner ones.
  - The ends of cables may show signs of relative movement between the cable and the socket or end connection.
  - Corrosion is more likely to occur at the lower end of a cable as it tends to remain wet due to rainwater.
  - There is often a groove at the cable/socket junction where water can collect.
  - If the outside of a cable is painted, the presence of a broken wire in the outer layer is indicated by a spiral crack in the paint caused by the relative movement between adjacent wires when the tension in the broken wire is released.
  - Areas where the lay of the cable or individual strands are disturbed are potential failure sites.
  - Potential corrosion of cables in post-tensioned precast concrete units.

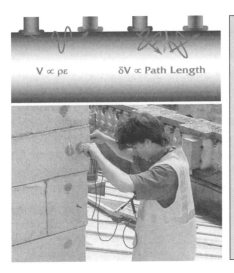

$$V \propto \rho\varepsilon \qquad \delta V \propto \text{Path Length}$$

Crack detection

UPV crack detection to determine weathering depth in limestone. An acoustic wave should pass through solid undifferentiated material at a constant speed that is proportional to the density and elastic modulus of the material. Cracks and discontinuities alter the path of the waves, extending the travel time. An open crack deflects the path as acoustic energy will not couple across the gap. Weak weathered limestones slow the waves from about 4 km/s to <1.5 km/s, depending on the severity of the weathering. (Courtesy GB Geotechnics Ltd., Cambridge.)

3.2.2.5 Plastics

The types of plastics used in construction include; polyvinylchloride, acrylonitrile butadiene styrene, acrylics, polystyrene, polypropylene, polyesters and epoxides. Plastics may be degraded in use by ultraviolet and infrared radiation, moisture etc. As a result crazing, resin-glass interface failure, colour fading and subsequent diminution of structural properties may occur. Rubber is regarded as a natural polymer.

For an experienced practitioner it may be possible to identify one or other of the above plastics by appearance, bounce, odour, feel, colour, specific gravity and the Beilstein test which detects halogens. Simple heating tests can determine the class of plastic and are usually carried out before an elemental test for nitrogen, sulphur, chlorine, bromine and fluorine. Infrared spectrographic analysis can also be used to identify plastics and their degradation products.

Other materials

Tests for other materials include that to check the chemical constituents of glass. (BS2649-1:1988(1993), *Glasses of the soda-lime-magnesia-silica type.*)

3.3 Monitoring

There is a popular truism in the construction industry that if you cannot measure it you cannot manage it. A developing aid to managing a situation is the employment of monitoring techniques. Monitoring to check for structural performance, deterioration and other parameters has advanced considerably in recent times. Available techniques include:

- Structural changes, material degradation, reinforcement corrosion, moisture movements and thermal response.
- Measurement of water content in masonry.
- Buildings affected by foundation heave, subsidence and lateral movement.
- Tunnel excavations constructed using both the New Austrian Tunnelling Method (NATM) and traditional boring techniques.
- Clay soils susceptible to shrinkage and expansion.
- Buildings threatened by collapse of old mine workings or hill-slip.
- Movements and vibrations caused by nearby construction.
- Vibrations and noise due to heavy machinery affecting the performance of a specialist section of a building such as a hospital operating theatre.
- Internal atmospheric conditions leading to higher than expected energy costs or staff complaints.
- The development of cracking in masonry and other materials aided by the use of crack monitors or vernier gauges.
- Use of anemometers to check wind speeds in and around structures.

As a recent example BRE were commissioned to monitor the Mansion House in the City of London as it was affected by movement during construction of a tunnel for the Docklands Light Railway (DLR). A comprehensive monitoring system including electro-level tilt meters and water level gauges to measure differential vertical movement, invar-wire extensometers to determine wall movement, load cells to monitor tie bar loads and thermistors to monitor temperature. The monitoring systems transferred processed data electronically to consultants' offices so that tunnelling work could be controlled hour by hour.

Similar arrangements were in place during construction of the Jubilee Line for London Underground Ltd. (LUL).

In many cases it may be necessary and helpful to monitor the performance of a building or structure before proceeding with refurbishment or repair work.

3.4 Sources of information

3.4.1 Institutions

Several professional institutions make available an on-line library catalogue. This is usually available free to members and may be free to non-members. Most institutions charge a fee for photocopying and for postage on documentation made available by fax or post.

Royal Institute of British Architects (RIBA)

There are in excess of 28,000 registered architects and their contact details may be accessed by using the RIBA website.

The Royal Institute of British Architects runs the RIBA Library in London and the collections and services are available to all. There is currently a day ticket fee of

£10 for non-members to use the reference library. The Library catalogue enables searching by building and architect name. The RIBA have an extensive on-line catalogue and also a Drawings and Archives Collections located at the Victoria and Albert Museum. There is also an on-line service to gain information on *Great Buildings around the World*.

Institution of Civil Engineers (ICE)

The ICE maintains one of the largest and most comprehensive libraries in the world (over 100,000 titles searchable on-line). Also of particular interest to practitioners will be the Concrete Archive of Reinforced and Prestressed Concrete. This is sponsored by ICE and the Concrete Society and was set up after an exhibition in 1996. Its aim is to record practice from the 1890s to the present day. It contains records of drawings, photographs and calculations of historic reinforced concrete (RC) structures and also information of proprietary systems. Most technical papers published in ICE Journals are available on-line at a modest cost.

In addition the ICE maintains an electronic civil engineering exchange with an email address (see Appendix B) which allows an electronic interchange of ideas and information. This system is freely available to members and non-members.

The virtual library provides on-line answers to 200 questions on practical civil engineering works under the following basic headings:

- bridge works
- concrete structures
- drainage works
- earthworks
- piers and marine structures
- roadworks
- pumping stations
- reclamation
- water retaining structures and waterworks
- pipe jacking and microtunnelling
- piles and foundations
- general issues

Institution of Structural Engineers (IStructE)

The IStructE maintains a comprehensive library of in excess of 42,000 titles, including documents that date back to the 1700s. The catalogue is searchable, for all, on-line. It also publishes a range of technical reports on topics which include:

- bridges
- appraisal of structures
- cladding
- deep basements
- car parks

- underwater structures
- adhesives
- surveys and inspection of buildings
- fibre composite reinforcement for concrete
- design of masonry, reinforced concrete, steel
- alkali-silica reaction
- structural failures
- glass
- subsidence
- access gantries

Of particular interest to practitioners are Informal Study Groups (ISGs) dealing with, for example, the history of structural engineering, heritage structures, management and maintenance of bridges, fire Engineering and arch bridges.

All technical papers published in *The Structural Engineer* are available on-line

Association of Building Engineers (ABE)

The ABE (formerly The Incorporated Association of Architects and Surveyors and founded by Sir Edward Lutyens in 1926) is a multidisciplinary body of construction professionals specialising in the technology of building and the built environment. The Association provides the qualification Building Engineer which is recognised internationally in both the public and private sectors. It is represented on codes, standards, legislative, technical and other advisory bodies. It gives advice on all matters related to construction and the built environment.

Royal Institution of Chartered Surveyors (RICS)

The RICS is particularly strong in commercial information such as guide costs of buildings and trends in house prices by regions. It maintains a technical library in London and Edinburgh. There is an entrance fee for non-members. The organisation also provides an online catalogue.

Chartered Institute of Building Services Engineers (CIBSE)

CIBSE maintains an on-line bookshop offering its own publications (see below for list of topics) and a selection of others. CIBSE publications include those on:

- Commissioning
- Lighting
- Heating, air conditioning and refrigeration
- Ventilation and indoor air quality
- Public health engineering
- Energy, sustainability and the environment
- Controls
- Electrical services
- Energy and environmental modelling

- Lifts and escalators
- Fire safety
- Facilities management and maintenance
- Project management
- Building services

Chartered Institute of Building (CIOB)

Based in Ascot, where it maintains a technical library. It claims to have the world's largest collection of construction literature. The on-line catalogue is available to all.

The British Library (BL)

Located at St Pancras, Boston Spa and Colindale, the BL library collection catalogue is searchable on-line free of charge.

Historic Scotland (HS)

HS is an executive agency of the Scottish Government. Its aim is to safeguard Scotland's historic environment while promoting its enjoyment by the public. It can schedule sites and list buildings and is responsible for the care of many of Scotland's historic sites.

The organisation provides advice notes on a number of topics including:
- Preparation and use of Lime Mortars (1995)
- Conservation of plasterwork (1994)
- Performance standards for timber sash and case windows (1994)
- Thatch and thatching techniques (1996)
- The Hebridean Blackhouse (1996)
- Earth structures and construction in Scotland (1996)
- Access to the built heritage (1996)
- Historic Scotland guide to international conservation charters (1997)
- Stone cleaning of granite buildings (1997)
- Biological growths on sandstone buildings (1997)
- Fire protection measures in Scottish historic buildings (1997)
- Quarries of Scotland (1997)
- The archaeology of Scottish thatch (1998)
- The installation of sprinkler systems in historic buildings (1998)
- Lime harling and rendering (2000)
- Burrowing animals and archaeology (1999)
- Bracken and archaeology (1999)
- The treatment of graffiti on historic surfaces (1999)
- Scottish aggregates for building conservation (1999)
- Corrosion in masonry clad early 20th century steel framed buildings (2000)
- Scottish slate quarries (2000)
- Fire risk management in heritage buildings (2001)
- Non-destructive investigation of standing structures (2001)

Guides for Practitioners:

- Stone cleaning – a guide for practitioners (1994)
- Timber decay in buildings – the conservation approach to treatment (1999)
- Rural buildings of the Lothians: conservation and conversion (1999)
- Conservation of historic graveyards (2001)

English Heritage (EH)

EH is the national body for the conservation of the built heritage in England. It provides grants, technical advice and guidance as well as owning over 400 historic properties, all open to the public. EH has a statutory role in determining the applications affecting the demolition of buildings which are listed or in conservation areas, and the alteration of nationally important listed buildings (see also Chapter 6). Typical EH publications include:

- *Research and Professional Guideline No.3: Scaffolding and temporary works for historic buildings*
- Notes on the conservation of historic bridges and on historic bridge parapets

Society for the Protection of Ancient Buildings (SPAB)

Publishes a number of advisory monographs of particular interest to practitioners dealing with old buildings. Topics include:

- Repointing stone and brick walling
- Control of damp in old buildings
- Care and repair of thatched roofs
- Panel in-filling to timber-framed buildings
- Repair of timber-framed buildings
- Timber bellframes
- Care and repair of old floors
- Care and repair of flint walls
- VAT and historic buildings
- Historic buildings: controls and grants
- A stitch in time: maintenance guide
- Basic lime-wash
- Timber treatment: defrassing and surface treatment
- The need for old buildings to breathe
- Removing paint from old buildings
- First aid repair to traditional farm buildings
- Tuck pointing in practice
- An introduction to building limes
- Patching old floor boards
- Rough-cast in old buildings
- Introduction to the repair of lime-wash and plaster

- How to make beeswax polish
- Is timber treatment always necessary?

3.4.2 Museums

The Royal Engineers Museum

Based in Gillingham, Kent, this museum and library is a useful contact point for all military construction, including, for example, information about Bailey bridging.

Other museums, such as the London Museum, hold information concerning the construction of local historic buildings.

3.4.3 Authorities

National House Building Council (NHBC)

The NHBC essentially provides latent defect insurance cover for housing. It does, however, support a range of technical literature defining its own required standards. These include a useful booklet entitled *Standards for conversions and renovations* which is available free of charge.

Construction Industry Council (CIC)

The CIC is the representative forum for the professional bodies, research organisations and specialist business associations in the construction industry. In addition CIC represents the views of the higher levels of the industry (professional, managerial and technical) in ConstructionSkills – the Sector Skills Council for Construction. Construction Skills is a partnership between CIC, CITB-Construction Skills and CITB Northern Ireland.

Standing Committee on Structural Safety (SCOSS)

This committee is supported by IStructE (which provides the secretariat) and ICE and was originally formed in 1976. Its terms of reference are to:

- consider both current practice and likely development from the standpoint of structural safety;
- be aware of trends and innovations in design, construction and maintenance from the point of view of safety;
- consider whether unacceptable risks exist or might arise in the future and, if believed so, to give warning to relevant bodies;
- consider whether further research and development appears desirable from the standpoint of structural safety;
- disseminate the findings of the Committee in a biennial published report and by other appropriate means;

- avoid duplicating the work of the Health and Safety Executive, of the Institution of Civil Engineers and of the Institution of Structural Engineers; and
- report to the presidents of the Civil and the Structural Engineers annually and from time to time on specific issues.

SCOSS has reported on a variety of topics including:

- Bridges under pedestrian loading
- Cast iron columns
- Change of use and higher loadings
- Competence and integrity
- Confidential Reporting on Structural Safety (CROSS)
- ISO 9001 definition of a 'product' of a design consultancy
- Education of engineers
- Effects of climate change
- External tubular structures – damage due to water ingress
- Glass balustrades
- Government proposals on reforming the law on involuntary manslaughter
- HSE discussion document: *Reducing Risks, Protecting People*
- Inspection of structures
- Management of structural reliability
- Microbiological attack on concrete
- Microbiological attack on steel
- NATM tunnels collapse, Heathrow Airport
- Overhead non-structural glazing
- Plywood beams supporting flat roofs
- Regular inspection of buildings
- Report of the Study Group on Structural Codes in Construction
- Rising groundwater
- Risk assessment
- Self certification and independent checking
- Stability of terraced housing
- Sports stadia – dynamic excitation of cantilever structures
- Supervision and checking
- Sustainable development
- Temporary condition of bridges
- Timber balconies
- Timber trussed rafter roofs in fire
- Tunnel linings and fires
- Underpinning
- Unplanned collapses during demolition
- Walkway collapse e.g. Ramsgate
- Warnings

Architects Registration Board (ARB)

(Formerly ARCUK – Architects Registration Council of the United Kingdom.) Responsible for the registration of architects in the UK, ARB may be able to assist with information about architectural practices that have been taken over or subsumed into larger concerns.

Association for Consultancy and Engineering (ACE)

ACE represents the business interests of the consultancy and engineering industry in the UK, and embraces around 800 firms operating over a number of disciplines. This organisation should be able to assist with contact details for firms within its purview.

Environment Agency (EA)

The EA deals with the environment, air quality, conservation, waste, water quality and resources, pollution prevention and control, radioactivity, agriculture and many other aspects of daily life.

Local Authorities

The make up of local authorities varies but most will have departments dealing with topics such as Building Control, Planning, Eduction and Roads, and will be able to respond to requests for information.

London District Surveyors Association (LDSA)

The LDSA was originally formed in 1845 and now represents local authority building control in the Greater London area. This organisation may be helpful in advising practitioners about general matters relevant to building control, particularly on health, safety and welfare.

Public Utilities

Such as, for example, Transco, British Telecoms, British Gas, Water Companies.

National Joint Utilities Group (NJUG)

This group maintains a protocol for the correct placement of underground services.

London Underground Ltd (LUL)

LUL is the controlling organisation for the London Underground and should be able to guide practitioners to the appropriate source on engineering and architectural matters.

Network Rail (formerly Railtrack)

Network Rail owns and operates the UK rail infrastructure and should be used as a first point of contact over engineering and architectural matters.

Port of London and other Marine Authorities

(*Dock Engineering* 3rd Edn (Brysson, Cunningham, 1922, Griffiths and Co, London.) provides details of many of the major docks and harbours built in ancient times in both the UK and abroad.)

Water Research Group (WRc)

WRc is an innovative, research-based group providing consultancy in the water and environmental sectors. It assists government and regulatory bodies in creating regulations soundly based on expertise obtained through 80 years of experience.

Ordnance Survey

Ordnance Survey maps are updated on a regular basis and comparison between updates is helpful in tracking the development of sites.

3.4.4 Trade Associations

- Steel Construction Institute (SCI)
- British Constructional Steelwork Association (BCSA)

(Of particular interest is the publication *Historical Structural Steelwork Handbook* (Bates 1984))

- Concrete Society (CS)
- Concrete Centre
- Brick Development Association (BDA)
- Timber Research and Development Association (TRADA)
- Rubber and Plastics Research Association (RAPRA)
- Glass and Glazing Federation (GGF)
- Lead Development Association (LDA)

Bibliography and further reference

Notes:

(1) For a further bibliography, see the IStructE Report *Appraisal of existing structures*, 2nd edn.

(2) Further reading is to be found in the Ashgate series on the history of civil engineering.

General

Allinson, K. (Ed.) 2006. *London's contemporary architecture*, Architectural Press. Oxford.

Armer, G.S.T. 2001. *Monitoring and assessment of structures*, E&FN Spon, London.

Ballard, G.,1999. *Look before you leap*, Building Conservation Directory (BCD), Wiltshire. Special Report on Ecclesiastical buildings.

Barton, N. 2005. *The lost rivers of London*, Historical Publications, London.

Beckmann, P. and Bowles, R. 2004. *Structural aspects of building conservation*, 2nd edn., Elsevier, Oxford.

Blanc, A. McEvoy, M. and Plank, R. (Eds.) 1993. *Architecture and construction in steel*, E & FN Spon, London

Brand, S. 2007. *How buildings learn: what happens after they're built*, 2nd edn., Phoenix Illustrated, London.

Brunskill, R.W. 1994. *Timber building in Britain*, 2nd edn., Cassell, London.

BRE. 1989. *Simple measuring and monitoring in low rise buildings: Part 1 cracks*, Digest 343, BRE, London.

BRE. 1989. *Simple measuring and monitoring in low-rise buildings: Part 2 settlement, heave and out-of-plumb*, Digest 344, BRE, London.

Buchanan, A. and Hudson, W.H. 1926. *Building construction plates*, Parts 1 and 2, BT Batsford, London.

Chrimes, M. 2006. Historical research: a guide for civil engineers, *Proceedings of the ICE, Civil Engineering*, 159(1), Thomas Telford, London.

Doran, D.K. and Cockerton, C. (Eds.) 2006. *Principles and practice of testing in construction*, Whittles Publishing, Scotland.

Foster, J.S. 1963. *Mitchell's advanced building construction*, 17th edn., BT Batsford, London. This book was first produced in 1893 and earlier editions are available in reputable libraries. They can assist in the understanding of early construction techniques.

Freitag, J.K. 1903. *The fire proofing of steel buildings*, Wiley, New York.

Friedman, D. 1995. *Historical building construction: design, materials and technology*, W.W. Norton and Co., New York.

Guedes, P. (Ed.) 1979. *Architectural and technological change*, Macmillan, Basingstoke.

Hewitt, C.A. 1980. *English historic carpentry*, Phillimore, Chichester.

Heyman, J. 1997. *The stone skeleton: structural engineering of masonry architecture*, Cambridge University Press, Cambridge.

Historic Scotland. 2001. *Non-destructive investigation of standing structures: Technical Advice Note 23*, Historic Scotland, Edinburgh.

IStructE. 1996. *Appraisal of existing structures*, 3rd edn. IStructE, London. When appraising existing structures recourse to the 1st edition (1980) and 2nd edition (1996) may also be helpful.

IStructE. 1997. *Structural assessment: the role of large and full scale testing*, IStructE, London.

Mainstone, R.J. 1975. *Developments in structural form*, Allen Lane, London.

McKay, J.K. 1943. *Building construction, Vol 1*, Longmans, Harlow.

McKay, J.K. 1944. *Building construction, Vol 2*, Longmans, Harlow.

McKay, J.K. 1944. *Building construction, Vol 3*, Longmans, Harlow.

McKay, J.K. 1961. *Building construction, Vol 4*, Longmans, Harlow. Several editions of the McKay series have been produced including those using the metric system.

Middleton, G.A.T. (Ed.) c.1901. *Modern buildings their planning, construction and equipment*, Caxton Publishing, London.

Moss, R.M. and Matthews, S.L. 1995. In-service monitoring: a state of the art review, *The Structural Engineer*, 73(2), IStructE, London.

Noy, E.A. and Douglas, J. 2005. *Building surveys and reports*, 3rd edn., Blackwell, Oxford.

Perry, P. and Thomas, R. 2009. Drawings, records and the quest for information on an existing building, *The Structural Engineer* 87(4), IStructE, London.

Robson, P. 2005. *Structural appraisal of traditional buildings*, Donhead Publishing, Shaftesbury.

Smith, D. (Ed.) 2001. *Civil Engineering Heritage: London and Thames Valley,* Thomas Telford, London.

Smith, S. 2005. *Underground London,* Little Brown, London.

Concrete

BCA. 1988. *The diagnosis of alkali-silica reaction,* British Cement Association, Surrey. (See also Chapter 5 in this book.)

BRE. 1981. *Carbonation of concrete made with dense material aggregates,* Information Paper IP6/81, BRE, London.

BRE. 1980. *Internal fracture testing of insitu concrete – a method of assessing compressive strength,* Information Paper IP22/80, BRE, London.

BRE. 1986. *Determination of the chloride and cements of hardened concrete,* Information Paper *IP21/86,* BRE, London.

BS 1881-122: 1983. *Testing of concrete. Method for determination of water absorption,* British Standards Institution, London.

BS 1881-202: 1986. Testing concrete, BSI, London. This standard is presented in several parts dealing with testing of hardened concrete for properties other than strength, density, tensile splitting strength, static modulus of elasticity, guidance on NDT methods, surface hardness by rebound hammer, measurements of velocity of ultrasound pulses, use of radiography and near-to-surface tests.

BS 1881-203: 1986. *Testing concrete – recommendations for measurement of velocity of ultrasound pulses in concrete,* BSI, London.

BS 1881-207: 1992. *Recommendations for the assessment of concrete strength near-to surface tests,* BSI, London.

BS 4550-0: 1978. *Methods of testing concrete – general introduction,* British Standards Institution, London.

BS 4550-2: 1970. *Chemical tests,* BSI, London.

BS 4551: 1980. *Methods of testing mortars, screeds and plasters,* BSI, London.

BS 6089: 1981. *Guide to assessment of concrete strength in existing structures,* BSI, London.

Bungey, J.H. 1992. *Testing of concrete in structures – a guide to equipment for testing concrete in structure,* Technical Note 143, CIRIA, London.

CS. 1987. *Concrete core testing for strength.* Technical report 11, The Concrete Society, Surrey.

CS. 1988. *Permeability testing of site concrete, a review of methods and experience,* Technical Report 31, The Concrete Society, Surrey.

CS. 1989. *Analysis of hardened concrete,* Technical Report 32, The Concrete Society, Surrey.

ICE. 1970. *Symposium on non-destructive testing of concrete and timber London 1969,* ICE, London.

Somerville, G. 1986. The design life of concrete structures, *The Structural Engineer,* 64(2), IStructE, London.

Stain, R.T. and Dixon, S. 1993. Inspection of cables in post-tensioning bridges – what techniques are available?, *Structural faults and repair,* 93(1), University of Edinburgh Press, Edinburgh.

Timber

Baird, J.A. and Ozelton, E.C. 1984. *Timber designers manual,* 2nd edn., Granada Books. London.

Bravery, A., Berry, R., Carey, J. and Cooper, D. 2003. *Recognising wood rot and insect damage in buildings,* 3rd edn., Report BR453, BRE Press, Garston, UK.

BRE. 1975. *The durability of glues used for plywood manufacture,* BRE, Garston, UK.

BRE. 1984. *Ageing of wood adhesives – loss of strength with time,* Information Paper 8/84, BRE, London.

BRE. 1992. Damage by wood-boring insects, Digest 307, BRE, Garston, UK.

BS 4978: 1988. *Specification for softwood grades for structural use,* BSI, London.

BS 5268-2: 1991. *Structural use of timber. CP for permissible stress design, materials and workmanship,* BSI, London.

BS 5666-2: 1980. *Methods of analysis of wood preservatives and treated timber,* BSI, London.

TRADA. 2006. *Moisture in timber,* Wood Information Sheets 4–14. Timber Research and Development Association, High Wycombe, UK.

TRADA. 1985. *Design stresses for members graded in-situ for British grown hardwoods,* Timber Research and Development Association, High Wycombe, UK.

Masonry

Berger, F. 1986. *Zur nachtraglichen Bestimmung der Tragfahigkeit von zentrisch gedrucktem Ziegelmauerwerk,* Ernst & Sohn, Berlin.

BRE. 1995. *Masonry and concrete structures: measuring insitu strength and elasticity using flat jacks,* Digest 409, BRE, Garston, UK.

De Vekey, R.C. 1994. The non-destructive evaluation of masonry materials in structures, *Proc. 8th CIMTEC World Conference,* CIMTEC, Italy.

De Vekey, R.C. 1990. *Corrosion of steel wall ties – recognition and inspection,* Information Paper P13/90, BRE, Garston, UK.

De Vekey, R.C. 1995. Thin stainless steel flat-jacks: calibration and trials for measurement of in-situ stress and elasticity of masonry, *Proc. 7th Canadian Masonry Symposium,* Hamilton.

Edgell, G. (Ed.) 2005. *Testing of Ceramics in Construction,* Whittles Publishing, Scotland.

Ferguson, W.A. *et al.* 1994. The screw pull-out tests for the in-situ measurement of the strength of masonry materials. In: *Proceedings of the 10th International Brick and Block Masonry Conference,* University of Calgary, Canada, July 1994.

RILEM. 1994. LUM D1 Removal and testing of specimens from existing masonry. In: RILEM *Recommendations for the Testing and Use of Constructions Materials,* E & FN Spon, London.

Metals

Ashurst, J. and Ashurst, N. 1988. *Practical building conservation: Vol 4: Metals,* English Heritage Technical Handbook, Gower Technical Press, Aldershot, England.

Blanchard, J., Bussell, M. and Marsden, A. 1982, Appraisal of existing ferrous metal structures, *The Arup Journal,* 18(1), ARUP, London.

BS 240: 1986. *Method for Brinell hardness test for verification of Brinell hardness testing machines,* BSI, London.

BS 427: *1990. Method for Vickers hardness test and for verification of Vickers hardness testing machines.* BSI, London.

BS 860: 1967. *Tables for comparison of hardness scales,* BSI, London.

BS 891: 1989. *Method for hardness tests (Rockwell method) and for verification of hardness testing machines,* BSI, London.

BS 2600-1: 1983 and BS 2600-2: 1973. *Radiographic examination of fusion welded butt joints in steel,* BSI, London.

BS 4080-1: 1989. *Surface discontinuities revealed by magnetic particle flaw detection,* BSI, London.

BS 4080-2: 1989. *Specification for severity levels for discontinuities in steel castings, Surface discontinuities revealed by penetrant flaw protection,* BSI, London.

BS 4124: 1991. *Methods for ultrasonic detection of imperfections in steel forgings,* BSI, London.

BS 417: 1989. *Method for superficial hardness test (Rockwell method) and for verification of superficial hardness testing machines (Rockwell method),* BSI, London.

BS 5289: 1976. *Code of practice for visual inspection of fusion welded joints,* BSI, London.

BS 6072: 1981. *Method for magnetic particle flaw detection,* BSI, London.

BS 6208: 1990. *Methods for ultrasonic testing of ferritic steel castings including quality levels,* BSI, London.

BS 6533: 1984. *Guide to macroscopic examination of steel castings including quality levels,* BSI, London.

BS 7448-1: 1991. *Fracture mechanics toughness tests,* BSI, London.

BS EN 10002-1: 1990. *Tensile testing of metallic materials. Method of tests at ambient temperature,* BSI, London.

BS EN 10002-5, 1992. *Tensile testing of metallic materials. Method of tests at elevated temperature,* BSI, London.

Doran, K. (Ed.) 1992. *Construction materials reference book,* Butterworth-Heinemann, Oxford. (A second edition is in preparation.)

Käpplein, R. 1991. Zur Beurteilung des Tragverhaltens alter gusseisener Hohlsäulen, *Berichte der Versuchsanstalt für Stahl, Holz und Steine der Universität Fridericiana in Karlsruhe,* 4(23), Karlsruhe, Germany. (In German: Ultrasonic tests for flaws and strengths: cylinder-splitting tests compared with tensile tests.)

Lazar, H.L. 1966. *Stress corrosion of metals,* Wiley, New York.

PD 6513: 1985. *Magnetic particle flaw detection – a guide to the principles and practice of applying magnetic particle flaw detection in accordance with BS 6072,* BSI, London.

Taylor, J.L. 1989. *Basic metallurgy for non-destructive testing,* British Institute of Non-Destructive Testing, Northhampton.

Plastics

Doran, K. (Ed.) 1992. *Construction materials reference book,* Butterworth-Heinemann, Oxford. (A second edition is in preparation.)

Lyons, A. 2007. *Materials for Architects and Builders,* 3rd edn., Butterworth-Heinemann, Oxford.

4 Types of contract

4.1 General

A number of organisations take an interest in contracts and methods of working. These include:

- The Construction Industry Council (CIC);
- The British Property Federation (BPF);
- Joint Contracts Tribunal (JCT);
- Fédération Internationale des Ingénieurs-Conseils (FIDIC); and
- Professional Institutions, such as RIBA, ICE, IChemE, RICS and IET.

There is no ideal type of contract for refurbishment or repair contracts. Many types of contract are available, some of which are tailor-made for a particular project. There has, however, been a move since the late 1960s to standardise forms of contract and below are brief descriptions of some of those standard forms. One important benefit claimed for standard contracts is that they spread the risks clearly and equitably between the parties. However some practitioners still take the view that this situation has not been achieved. In a recent, well reasoned, article in *The Structural Engineer* (TSE), John Carpenter (Secretary to SCOSS) argues strongly for adequate interaction between constructors and designers. He contrasts the better practice in other industries to that practised by construction.

Practitioners should be aware that under EU rules (for publicly funded contracts) a threshold has been established beyond which it is mandatory to invite tenders for construction work to all members. At the time of writing these were:

- €5,271,352 (£3,611,474) for works; and
- €210,854 (£144,459) for supplies and services, including consultancy fees.

Prudence suggests a check on these levels before submitting projects to tender. Up to date details may be obtained from the Office of Government Commerce (OGC) website.

In recent times a number of changes have contrived to stress the importance of fair and tight contracts. These include the:
- crucial interaction between the law of contract, tort and restitution demanded by injured parties;
- relationship between common law and standard forms of contract; and
- introduction of the *Housing Grants and Construction and Regeneration Act 1996 Part II.*

Concerns shown by the industry are illustrated in a recent Steel Industry Guidance Note issued jointly by CORUS, the British Constructional Steelwork Association (BCSA) and the Steel Construction Institute (SCI) which, in a slightly edited version, poses the following questions:

- When does payment actually start?
- Does the payment cycle start when work is undertaken off-site or does it begin when work has started on site?

But other questions also arises:

- Is there something in the contract conditions that delays the start of the payment cycle? More and more frequently clauses appear providing that payment will not start until all the types of security required – for instance performance bond, parent company guarantee, warranties – have been issued and delivered.
- What are the payment terms? This is not always the simple question it seems. It is not unusual to find amendments to amendments to amendments; supplementary conditions, schedules and annexes; orders, numbered documents and minutes, all saying something about what must be paid and when (and more to the point, what is not to be paid or not paid yet). All of these have to be read in the context of the *Construction Act*, if it applies. It is sometimes next to impossible to trace exactly what the payment terms actually are.

The 'what' and 'when' of payment need to be clarified early as they will have a real impact on cash flow.

- What does the price cover? Surely the price covers what has been quoted for? Not always. For instance, the contract may require the subcontractor to check the work of other subcontractors. It is not reasonable to expect companies to have the requisite skill to do this, let alone to have included for it in the quotation.
- Are retentions to be held? Retentions are still the bane of the supply chain's life. The first half of retention is usually paid to the contractor upon practical completion. The second half is more problematical as it is commonly not paid until completion of the whole project (this may include subsequent defects found during the maintenance period). For contractors early on site, this may be some years ahead. Also, as payment is tied to completion of the whole project, problems down the line may prevent any of the retention being paid. It is also worth checking whether the retention is held on a 'pay when paid' basis, which is not enforceable on contracts covered by the *Construction Act*. Many enlightened clients are moving away from the use of retentions in their projects and in these cases, no retentions should be held in any part of the supply chain.

Alternatively it is worth offering some form of security in place of cash retention.

- Whose work am I liable for? You may find that you have made yourself liable for more than your own and your subcontractors' work. In the case of *The Co-operative Insurance Society Ltd v Henry Boot (Scotland) Ltd*, a contractor undertook the development of a consultant's design. Based on the wording of the contract the contractor was liable for the consultant's work.

- What does the contract allow as damages? In the event of a claim that arises for payment of damages, what does the contract allow? Some clauses only allow a subcontractor to recover what can be recovered by the main contractor under the main contract. In this case, a subcontractor cannot recover for loss caused by the default of the main contractor himself or of another subcontractor. As an exclusion clause, this is quite possibly not effective but who wants to have to go to court to find out?

- Are there indemnities in the contract conditions? An indemnity is an agreement to make good a loss suffered by another. The party indemnified does not have to take reasonable actions to minimise the loss (mitigate the loss) which would be necessary in ordinary contractual claims, neither does the loss have to be reasonably foreseeable. Additionally, indemnities can be long lasting. They can extend the usual 6–12-year limitation period during which you are at risk.

- Can the works be suspended for non-payment? Where a contract is covered by the *Construction Act*, there is a statutory right to suspend performance of obligations under it where a sum due is not paid in full by the final date for payment. The act requires at least seven days' notice to be given. Many contracts try to increase this period but arguably a longer period is not enforceable. Beware, however, that collateral warranties may also require longer notice periods and this will be enforceable. Two problems arise with suspension for non-payment: it can be tricky to be certain that a particular sum is due, and the consequences of getting it wrong could be catastrophic. Also, although the contract period will/may be extended by the length of the suspension, there is no automatic right to extra time for remobilisation (although some contracts allow this). This may change in the future.

- Costs in adjudication? There are two different things covered by this item – the adjudicator's own costs and the parties' costs. It is up to the adjudicator, subject to the law and applicable adjudication procedure, to decide what happens concerning adjudicator's costs. As far as the parties' costs are concerned, the courts have decided that – subject to anything to the contrary in the adjudication procedure – parties can agree that the adjudicator should be able to decide how costs are paid. That agreement may be very informal and *ad hoc*. Some contracts now require the party referring the dispute to adjudication to be responsible for all the costs of both parties, whatever the outcome of the adjudication. The referring party has no control over what costs the other party incurs. The Government has said it intends to outlaw this practice but it has not done so yet and for the time being, such clauses remain enforceable.

- What is the interest rate on overdue payments? The statutory rate is 8% above the Bank of England base lending rate, the idea being to discourage late payment rather than simply compensate for it. But the contract can vary the right to statutory interest provided the overall remedy for late payment is a *substantial* remedy. Check what rate you are being offered and challenge it if necessary. Don't forget that in addition to interest you are entitled to (small amounts of) compensation for late payment.
  - £40 for debts up to £999.99
  - £70 for debts between £1,000 and £9,999.99
  - £100 for debts of £10,000 or more

The document then summarises the concern of the steel industry in the following list of key points:

- Money is the most important issue in any contract.
- Provisions affecting money may appear throughout the contract, not just in the payment term.
- Many of these provisions may be unexpected.
- Terms are not always clear.
- Not all provisions will necessarily be enforceable at law but it is not wise to rely on this.
- Clients may accept security in lieu of retention.
- Indemnities can often be onerous.
- It may be possible to suspend work where payment has not been made on time.
- There is some good news as well in the form of statutory interest for late payment. The rate is set high to discourage late payment although this can be altered under some circumstances.

The reader may regard this as a plaintive *crie de coeur* written in forceful language by a disaffected section of industry, but it does graphically illustrate some of the difficulties that can be experienced when contracts are written in a biased manner. The note also indicates that there are residual difficulties with adjudication which was conceived as an additional and more efficient method of dispute resolution.

It is in an attempt to deal with questions of this type that standard contracts have been drawn up.

The construction industry has an unenviable reputation for dispute and much of the work of the above organisations is targeted towards dispute reduction. Many of the contracts sired or influenced by the above organisations contain specific dispute resolution techniques including negotiation, mediation, adjudication, arbitration and litigation. To these traditional methods have now been added expert determination and early neutral evaluation.

For repair contracts in response to insurance claims it is essential to avoid arguments in relation to betterment, which may arise when the repaired work is of a higher quality (or more extensive) than the work it replaces.

Many years of experience have indicated that successful contracts will only be achieved by adherence to the following principles:

- Not starting on site too early
- Having *all* details, schedules etc completed *before* site commencement
- Having *rigid control* of all post-contract variations. It sometimes pays to build the contract as originally detailed and then make variations *after* completion of the original contract.

4.2 Methods of working

The construction industry has an almost bewildering array of methods for working together. The principal participants are usually a client, a designer and a contractor. In some cases each of these components is a single entity; in others a duality of representation. For example for a simple house-building scheme (or refurbishment) the designer may be an architect who is a sole practitioner. For the re-vamp of a major industrial complex the designer may be a mechanical engineer supported by other disciplines which include civil/structural, architectural and building services. A main contractor may require the services of several specialist subcontractors. Below are briefly described some possible elements of working methods; in practice these may occur singly or in combination for any particular contract. For Public Finance Initiative (PFI) contracts there is a bias towards performance rather than prescriptive specifications in order to provide incentives to the private participants.

Major contracts for the refurbishment of a large university teaching hospital or school often require a partnership between public and private institutions. One way of providing this is through a Public Private Partnership (PPP). The PFI (Public Finance Initiative) is a well-developed type of PPP and is usually based around a tangible asset such as a building or group of buildings for which private finance is used to provide a facility and/or service in return for which the private sector receives payment over the life of the concession. In major schemes such as the redevelopment of the London Underground the life of the concession might be 25 or 30 years. Suspension or reduction of payment can be imposed as a penalty for poor or untimely performance on the part of private participants.

These arrangements currently find favour with government as they encourage integration of services and attention to design, procurement, construction, maintenance, sustainability and whole life costing. Additionally, and most importantly, the state is relieved of the responsibility of finding the capital for a scheme thus administered.

Whichever method of working is selected it is essential that the contract clearly delineates the duties and responsibilities of each party to that contract. It is worth remembering that the successful execution of a contract requires attention to the correct balance between Quality, Profitability, Programme and Safety. In connection with Programme it should be noted that many contracts commence on site too early and, as a result, suffer from a poor flow of technical information.

4.2.1 Traditional

In this method a client retains a consultant (frequently an architect but it could be an engineer or quantity surveyor) to act as his co-ordinator. The coordinator will organise the design, aided by specialist consultants (e.g. structural and services engineers), arrange for tender documents to be produced, tenders procured and a contractor selected. In fairness to all parties it is important to relate the number of contractors approached to the size and importance of the contract (usually in the range three to six). It is also important to allow contractors sufficient time to construct well thought out tenders. The co-ordinator will also act as contracts administrator. During construction he/she will play a role in achieving the specified quality of the work by supervising the contractor's Quality Plan. He/she may also act as Planning Supervisor or select another party to perform that duty. (Practitioners should note that, in 2007, the CDM regulations underwent a substantial modification, as a result of which the 'Planning Supervisor' has now become the 'Co-ordinator'. See Chapter 6 for further information.)

4.2.2 Design and build (D & B)

In this method of working, a contractor (selected either by competition or negotiation) will assume responsibility for both design and construction. As few contractors nowadays employ design staff, it is customary for a contractor to work with a design consultant in the tender phase. The successful contractor may then appoint the same consultant to draw up working details or may opt for an alternative. (The term used when the client or contractor appoints a designer who is then transferred to the successful contractor to complete the working details is 'novation'.) The popular view is that this procurement method gives a client cost security but possibly at the risk of an adequate but poorer quality design. However some practitioners take the view that any variations to scope can result in severe cost penalties. Variations and extensions of D & B include DBO (Design, Build and Operate) and DBFO (Design, Build, Finance and Operate).

4.2.3 Management contracting

In this system the client (e.g. developer) engages the contractor at an early stage. The contractor will not carry out the work but will manage the process. All work will be subcontracted to works and/or specialist subcontractors.

4.2.4 Construction management

In this form of arrangement the client enters into a direct contract with each specialist. These specialists may, for example, include contractors, architects, engineers, surveyors and independent test houses. The client employs the construction manager to act in a consultative capacity to co-ordinate the efforts of these specialists.

4.2.5 Turnkey

This is arguably the most comprehensive method of working, where one organisation procures all items needed by a client. Most commonly used for projects such as oil refineries where a major contractor might tender for site provision, design, construction, commissioning (including staff provision) and ongoing maintenance.

4.2.6 Partnering

A recently employed term which usually denotes a voluntary agreement between parties that regularly work together because of common expertise or even friendship. Consultants and contractors are sometimes prepared to work for little or no fee to secure a contract and then to enter into a formal agreement on securing the project. Experience suggests that to continue without a formal contract setting out the role and responsibilities of the partied is foolish and irresponsible. It is important that, when entering the contract stage, one person or organisation is nominated as leader of the team and given the ultimate responsibility for making major decisions. Most of the above types of contract may be used as a formal agreement for the parties to work together.

4.3 Standard types of contract

There follows brief notes on some of the standard types of contract available:

4.3.1 Joint Contracts Tribunal (JCT)

The Joint Contracts Tribunal (established in 1931) comprises representatives from ACE, British Property Federation (BPF), Local Government Association, National Specialist Contractors Council, Royal Institute of British Architects (RIBA), Royal Institution of Chartered Surveyors (RICS) and The Scottish Building Contract Committee. They have produced a number of standard contracts, the forms for which are available from publishers Sweet & Maxwell and the RIBA Bookshop. They are available in hard copy or CD.

In summary the JCT series covers the following:

- *Traditional. JCT SF 98 (+ CDPS): JCT IFC 98: JCT MW 98* (now obsolete)
- *Design and Build. 2005*
- *Management. JCT Management 98*

- *Minor Works 2005 + new MW with contractor's design*
- *Intermediate 2005 + new Intermediate with contractor's design*

4.3.2 New Engineering Contract (NEC)

The NEC was originally sponsored by the Institution of Civil Engineers (in particular by the Legal Affairs Committee) and the first edition proper appeared in 1993 under the guidance of Dr Martin Barnes. NEC2 was issued in 1995 and NEC3 in 2005 and was produced by the Institution of Civil Engineers through its NEC Panel.

The use of this series of documents was endorsed by Sir Michael Latham in his 1994 report *Constructing the Team*. NEC3 also comes with an endorsement from the Office of Government Commerce. It says that 'OGC advises public sector procurers that the form of contract used has to be selected according to the objectives of the project, aiming to satisfy the Achieving Excellence in Construction (AEC) principles. This edition of the NEC (NEC3) complies fully with the AEC principles. OGC recommends the use of NEC3 by public sector construction procurers on their construction projects.'

The NEC is a family of standard contracts and the latest contracts (June 2005) are:

- *NEC3 Engineering and Construction Contract*
- *NEC3 Engineering and Construction Short Contract*
- *NEC3 Term Service Contract*
- *NEC3 Professional Services Contract*
- *NEC3 Framework Contract*
- *NEC3 Engineering and Construction Subcontract*
- *NEC3 Engineering and Construction Short Subcontract*
- *NEC3 Adjudicator's Contract*

The Engineering and Construction Contract (ECC), which was previously the 'New Engineering Contract', is the most popular NEC contract used and has been developed to meet the current and future needs for a form of contract to be used in the engineering, building and construction industries. It is claimed to be an improvement on existing contracts in a number of ways including:

Flexibility

It can be used:

- For engineering and construction work containing any or all of the traditional disciplines such as civil, electrical, mechanical and building work.
- Whether the contractor has full design responsibility, some responsibility or no design responsibility.
- To provide all the current options for types of contract, whether negotiated or competitively tendered, such as lump sum (where the contractor

is committed to his offered prices), target cost, cost reimbursable or management contracts.

- In the United Kingdom and other countries.

Clarity and simplicity

Although a legal document, the ECC is written in ordinary language. As far as possible, it uses only words which are in common use. This makes it easier to understand by people who are not used to using formal contracts and by people whose first language is not English. It also makes it easier to translate it into other languages. However, in areas of insurance, disputes and termination, some phrases or terms which have a specific legal meaning have been retained. In particular the number of clauses used and the amount of text in each is less than in many standard forms; sentences are kept short and have been subdivided using bullet points to ease understanding; option clauses are designed so that they only add to the core clauses rather than alter or delete them. Finally the ECC neither requires nor contains cross-references between clauses.

Stimulus to good management

It is claimed that one of the most important characteristics of the ECC is that every procedure has been designed to contribute to, rather than detract from, the effectiveness of management of the project.

The twin principles upon which the ECC is claimed to have been based are:

- Foresight applied collaboratively mitigates problems and shrinks risk.
- Clear division of function and that responsibility helps accountability and motivates people to play their part.
- A prominent example of the way that the ECC is designed (it is claimed) to stimulate good management is the early warning procedure which is designed to ensure that the parties are made aware, as soon as possible, of any event which may:
  - Increase the amount that the Employer has to pay
  - Delay the performance of the works
  - Impair the performance of the works, once completed, or
  - Affect others working on the project

The ECC is presented on a 'Schedule of Options' basis as follows:

Core Clauses

General
The *Contractor*'s main responsibilities
Time
Testing and Defects
Payment
Compensation events

Title
Risks and insurance
Termination

Main Option Clauses

A. Priced contract with activity schedule
B. Priced contract with bill of quantities
C. Target contract with activity schedule
D. Target contract with bill of quantities
E. Cost reimbursable contract
F. Management contract

Dispute Resolution

Choice of two dispute resolution procedures, one that is compliant with The Housing Grants, Construction and Regeneration Act 1996 and one that is for use when this Act does not apply. Also has information on the appointment of an Adjudicator.

Secondary Option Clauses

Price adjustment for inflation; changes in the law; multiple currencies; parent company guarantee; sectional completion; bonus for early completion; delay damages; partnering; performance bond; advanced payment to the Contractor; limitation of the Contractor's liability for his design; retention; low performance damages; limitation of liability; Key Performance Indicators; the Housing Grants, Construction and Regeneration Act 1996 and the Contracts (Rights of Third Parties) Act 1999.

Schedule of Cost Components

Defines, *inter alia*, the make up of the wages or salary of the Contractor's directly employed workforce.

4.3.3 Institution of Civil Engineers (ICE)

These contracts have, to a large extent, been superseded by NEC3 but the following still exist and are available from ICE:

- *Conditions of Contract Target Cost Version,* 1st edn. (provided with guidance notes).
- *Conditions of Contract for Archaeological Investigation* (provided with guidance notes).
- *Conditions of Contract for Ground Investigation,* 2nd edn. (provided with guidance notes).
- ICE/ACE/CECA. *Design and construct conditions of contract,* 2nd edn., 2001.
- ICE/ACE/CECA. *Conditions of contract: measurement version.* 7th edn., 2003.

Fédération Internationale des Ingénieurs-Conseils (FIDIC)

The following contract forms are available through Cornerstone Seminars:

- *International Civil Engineering Conditions.* 4th edn. (1987/1992).
- *Construction Contract* (1999).
- *Plant and Design and Build* (1999).
- *EPC Turnkey* (1999).
- *Short Form* (1999).

UK Government

This long established list of contracts sponsored by HM Government are currently published by the Property Advisers to the Civil Estate and are stated to be *Latham* compliant and also in line with *The Housing Grants, Construction and Regeneration Act (1996)*

The following is a listing of contracts in the GC/Works portfolio first published in 1998. They are based on a series of contracts originally designed for Government contracts before the Second World War:

- *GC/Works/1: Part 1*—derived from foundation document. With quantities.
- *GC/Works/1: Part 2*—derived from GC/Works/1: Part 1. Varied without quantities.
- *GC/Works/1: Part 3*—derived from GC/Works/1: Part 1. Varied for design and build.
- *GC/Works/2*—derived from GC/Works/1: Part 2. Simplified version.
- *GC/Works/3*—derived from GC/Works/1: Part 1. Varied for mechanical and electrical works
- *GC/Works/4*—derived from GC/Works/2. Simplified version.

Others

Other standard forms of contract are available from the following institutions:

- Institution of Chemical Engineers (IChemE)
- Institution of Electrical Engineers (IEE)
- Royal Institute of British Architects (RIBA)
- Royal Institution of Chartered Surveyors (RICS)

Collateral warranties

A collateral warranty is a contract which gives a third party rights in an existing contract entered into by two separate parties. They exist due to the doctrine of 'privity of contract' i.e. only a party to a contract can benefit from it. The introduction of the *Contracts (Rights of third parties) Act 1999* was meant to do away with collateral warranties. However, this Act is contracted out of most of the major forms of contract, hence the remaining need for collateral warranties.

Practitioners involved in such matters are strongly advised to seek legal or other professional advice as the wording of such warranties can sometimes lead to unexpected results. In one notorious case, a supermarket had employed a developer to provide new facilities including a car park. Problems arose with the car park and when the developer went bankrupt the supermarket claimed for recompense against the contractor. There was a problem when it was revealed that the contractor was owed money by the developer!

The contractor mounted a defence against the claim, arguing successfully that this should be set off against the sum for which the supermarket was suing.

Hence collateral warranties should, in such a case, be amended to exclude the defence of 'set-off'.

Neil le Roux gives a lawyer's view on collateral warranties in Chapter 16 of Yu and Bull (2006).

Bibliography and further reference

Adriaanse, J. 2005. *Construction contract law: the essentials,* Palgrave Macmillan, Basingstoke.

BCSA *et al.* 2007. *Modern standard forms of contract: SN20 11/2007,* BCSA, London.

Bevan, A. 1992. *Alternative dispute resolution,* Sweet and Maxwell, London.

Campbell, P. (Ed.) 1997. *Construction disputes: avoidance and resolution,* Whittles Publishing, Scotland.

Carpenter, J. 2006. Designing for construction, *The Structural Engineer,* 84(10), IStructE, London.

Chappell, D., 2006, Construction contracts: Questions and answers, Taylor and Francis, Oxford.

Chappell, D., 2006, *The JCT Intermediate Contracts.* 3rd edn., Blackwell Publishing, Oxford.

CIRIA, 1994, *A Guide to the management of building refurbishment: Report 133,* CIRIA, London.

Gerrard, R. 2005. *NEC2 and NEC3 Compared,* Thomas Telford, London.

ICE and Totterdill B.W. 2006. *FIDIC User's Guide,* Thomas Telford, London.

ICE. 2006. *ICE Arbitration Procedure,* ICE, Thomas Telford, London.

ICE. 2006. *ICE Conditions of contract Target Cost version,* 1st edn., ICE, Thomas Telford, London.

ICE. 2006. *The Institution of Civil Engineers' Arbitration Procedure,* 3rd edn., ICE, Thomas Telford, London.

ICE and Gaitskell, R. (Ed.) 2006. *Engineers dispute resolution handbook,* Thomas Telford, London.

McLaughlin, R.T. P. and Doran, D.K. 1986. *Designer-contractor relationship,* IABSE, Zurich.

NEC. 2002. *The NEC Compared and Contrasted,* Thomas Telford, London.

Solt, G. and Hill, R. 2006. *Financial fundamentals for engineers,* Butterworth-Heinemann, Oxford.

Yu, C.W. and Bull, J.W. 2006. *Durability of materials and structures: in building and civil engineering,* Whittles Publishing, Scotland.

5 Types of construction: disasters, defects and potential solutions

What we anticipate seldom occurs; what we least expect generally happens.

– Benjamin Disraeli

5.1 General

Lawrance Hurst, writing in *Eminent Civil Engineers* (Doran 1991), wrote a piece entitled 'If you don't put a date to something you can't even be wrong'. He went on to say that the dating of buildings and bits of buildings is the best way to learn how building construction developed and the history of any particular building. This is excellent advice and is key to any investigation into existing construction. For example, it is unlikely that a substantial metal framed building built before 1900 will be of steel construction, although a British warship HMS *Iris* was built in steel in 1877. Construction dates are often displayed on foundation stones attached to buildings but care must be taken that these dates refer to the original construction and not just to a later addition (see Fig 5.1).

Figure 5.1 1889 building now converted into pharmacy (Courtesy Maureen Doran).

The construction industry has been slow to learn the lessons of the past; collective amnesia has been rife. Furthermore, clients are often reluctant to advertise mistakes so the results of forensic investigations remain locked in the filing cabinets

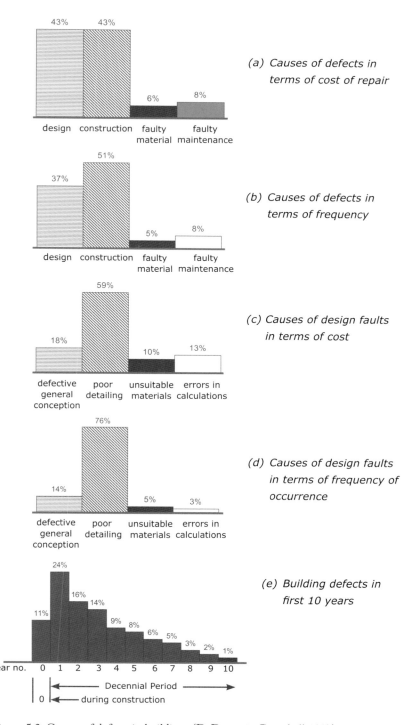

Figure 5.2 Causes of defects in buildings (D. Doran in Campbell 1997).

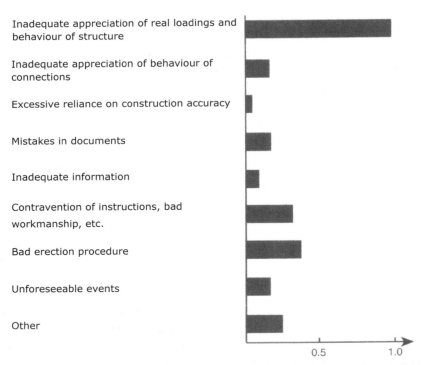

Figure 5.3 Prime causes of failure in 120 case histories (Poul Beckmann in Campbell 2001).

of lawyers. As a result many defects occur which should have been avoided. This chapter endeavours to illustrate some of those defects and suggest solutions. In a paper to the Institution of Structural Engineers in 1984 Dr Alistair Paterson analysed the origin of these faults and the results are shown in Fig. 5.2 (Paterson 1984, see also Campbell 1997). Although this analysis was based on statistics from France, the BRE at the time confirmed that British experience closely related to that of the French.

In a similar piece of research conducted by BRE and CIRIA between 1978 and 1980, Poul Beckmann reports (Campbell 2001) on 120 structural failures (see Fig. 5.3). It is claimed that this research differs from other similar work with regard to its methodology in the following respects:

- The method of survey was devised by a panel of professional engineers with considerable experience in design and construction practice and not only in investigating cases of failure.
- The survey information was provided by professional engineers with direct and detailed knowledge of the background to the cases they reported and of the circumstances in which the failure occurred.
- The information was provided on a strictly confidential basis and was processed in a non-attributable form, thus enabling the engineers to

make available all relevant information even though it may have been sensitive to claims and legal disputes. Such information would otherwise never have seen the light of day and its omission could have distorted the interpretation of the remainder of the information. A similar procedure was adopted by the Task Group which drafted the IStructE reports dealing with the structural aspects of alkali-silica reaction (ASR). In this case, bridge data was supplied by the DoT/Highways Agency (HA) and vetted by BRE. Coded summaries were then made available to the Task Group to assist with their work.

- Cases of loss of serviceability were reported as well as collapses. It was believed that in the past only a small proportion of cases of loss of serviceability had been studied with the result that comparative analysis was difficult or impossible.

In 1975 the BRE reported on failure patterns and implications from 510 investigations. Approximately half of these related to dampness. Other topics raised included faulty design, faulty materials and components, detachment (e.g. slip bricks) and unexpected user requirements.

Miroslav Matousek (1977), researching for the Swiss Federal Institute of Technology, analysed 800 construction failures and deduced that most were related to site malpractice; furthermore many could have been avoided without the introduction of demanding and bureaucratic checking procedures. Regrettably he detected over 500 fatalities and nearly 600 injuries in amongst those failures.

Although many defects are related to poor design and detailing, inadequate attention to the need for good workmanship has also been responsible for many problems. Unsuitable or faulty materials were only rarely the cause of failure.

In a paper to the Institution of Structural Engineers – *Construction safety: an agenda for the profession* – Dr Allan Mann has reiterated the need to learn from construction failures.

Section 5.1.1 lists significant incidents of interest to the construction industry for the period 1966–2005. Reviews of the type carried out by Dr Mann have been repeated from time to time but such failures continue to occur. It is time the industry did more to learn from its history.

5.1.1 Incidents in the period 1966–2005

Ferrybridge Power Station, 1965

Collapse of three natural draught, hyperbolic, reinforced concrete cooling towers in high winds. As a result, structural engineers were warned of the dangers of extrapolating design information on wind loading and also the effect on wind speeds of clusters of closely spaced structures. It is also of note that a further collapse took place of a Ferrybridge tower in 1984, due to the effect of comparatively small inaccuracies in construction having a disproportional effect on the strength of some structures.

Aberfan, 1966

Spoil tip failure resulting in collapse of school and many deaths and injuries. As a result the tips of many coal mines were reassessed for stability and those thought to be in risk of collapse lowered or repositioned. This incident brought about the need for a more scientific appraisal of spoil heaps and also a reconsideration of the siting of such disposal tips adjacent to vulnerable properties.

Sea Gem, 1967

Collapse of a steel offshore oil rig during movement to new site. Sea Gem, a jack-up barge employed on drilling in the North Sea, collapsed prior to relocation, with considerable loss of life. The public enquiry that followed revealed that the steel legs had a history of modification and were brittle at the prevailing temperature. There was also criticism that the mechanical, electrical and structural system had not been considered as a comprehensive unit.

Ronan Point, 1968

Progressive collapse of high rise block of flats following a moderate gas explosion (see Fig. 5.4). The block in question was 22 stories high and built using the precast concrete Larssen Nielsen system originally developed in Denmark in 1948. As a result of this collapse, there followed a major revision of the building regulations to safeguard against progressive and disproportionate collapse. In the aftermath all blocks over five stories built using large panel precast concrete panels were appraised for progressive collapse. As a result many blocks were strengthened using steel rolled

Figure 5.4 Ronan Point – partially collapsed building (J.C. Chapman in Campbell 2001).

steel angles to secure the joints between floor and supporting walls. This incident brought about a re-appraisal of the concept of disproportionate collapse and a modification of building regulations and design codes.

Emley Moor Mast, 1969

1265 ft high steel transmitting aerial near Huddersfield, Yorkshire collapsed as a result of ice accretion on the mast structures. As a result, more attention to ice accretion was adopted in new designs. It is generally acknowledged that some early guyed masts were built to support TV transmitting antennas and that their notional design life was 30 years. The need for an extended life has since arisen to cope with increased demand for support to additional TV antennas and other services. In order to verify continued structural performance some masts are now fully instrumented and continuously monitored. For further information on monitoring see Chapter 3. This incident brought about a re-appraisal of incidental loading to slender structures and their support systems in extreme conditions.

Bridges, 1970s

Problems (including collapses) with several box girder bridges (Avonmouth, Milford Haven, West Gate/Yarra) resulting in the promulgation of the Merrison Rules and subsequent inclusion in bridge standards and codes of practice.

Clarkson Toll, 1972

Gas explosion in Edinburgh in a large unventilated cavity, resulting in 21 deaths and many injuries. Although this incident highlighted the need for good ventilation of cavities and other similar structures it is not apparent that any modification of Building Regulations or the like occurred.

Sir John Cass School, 1973

A HAC problem resulting in the collapse of roof beams. As a result, the structural use of HAC was banned although is still used in refractory applications.

Summerland, Isle Of Man, 1973

Fire in an entertainment facility resulting in the loss of 50 lives. Rapid fire spread, flammable materials and poor means of escape criticised.

Rock Ferry School, 1976

Due to lack of robustness and poor support details, roof beams collapse.

Mersey House Bootle, 1976

Limited damage to high rise block of flats following moderate gas explosion of similar order of magnitude to that at Ronan Point. The structural design in this case featured an *in-situ* reinforced concrete frame which limited the collapse to the area immediately adjacent to the explosion. No loss of life was reported.

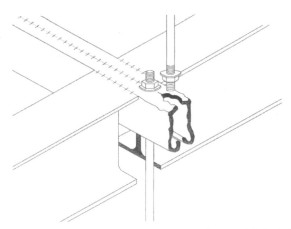

Figure 5.5 Sketch of the Hyatt Regency Hotel walkway failure, 17 July 1981, Kansas City, Missouri, USA (adapted from Feld and Carper, 1997) (Campbell 2001).

Hyatt Regency Hotel, 1981

Kansas City USA. Collapse of high-level walkways due to failure of steel support connection (see Fig. 5.5). A modification of original detail was requested by the steel fabricator and agreed by the original designer but not adequately checked. This resulted in 114 deaths and many serious injuries. This incident highlights the need to closely control and monitor any changes to details by fabricators and other subcontractors.

Abbeystead Pumping Station, 1984

Methane gas explosion in underground room killing several and injuring others. At the time the verdict was received with some scepticism in finding the consultant responsible for the disaster (see also Clarkson Toll above).

Carsington Dam, 1984

Partial collapse of dam structure. Earthwork slip during construction.

Bradford Fire, 1985

Fire in a football club grandstand resulting in loss of life and injuries. This incident highlights the need for practitioners to follow up designs and ensure that best practice is built into the way in which buildings and structures are used after completion.

Ynysygwas Bridge, West Glamorgan, 1985

Segmental, post-tensioned, I-beam construction bridge collapsed under its own weight and without warning (see also Piper's Row Car Park below).

The Great Storm in South East England, 1987

This was the worst to affect the south east of England since 1703. Some 15 million trees were felled and buildings suffered great damage. Sixteen people died as a result of the storm. Gust velocities of almost 100 mph were recorded at London Gatwick and in excess of 100 mph on the south coast. As a result, codes of practice dealing with the design of structures in wind were reconsidered. These figures were exceeded in January 2005 when gusts of 124 mph were recorded in Rona in the Western Isles, Scotland.

Piper Alpha oil platform 1988

Failure (explosions and a fire) of North Sea oil platform resulting in a large loss of life and many injuries. This incident highlights the need for good management systems in the use of complex plant.

King's Cross Underground, 1988

Wooden tread escalator fire resulting in 31 deaths and many injuries. Fire started when a discarded live match ignited grease and rubbish in a machine room. This incident highlights the need for the use of non-flammable materials and also good management systems in the every day use of facilities.

Hillsborough, 1989

Ninety-five people died and several hundred were injured due to overcrowding of football club grandstand. This event brought about a complete reassessment of the design of soccer stadia. It also brought into sharp focus the need to look carefully at the conditions pertaining when soccer and other sports stadia were used for alternative purposes such as pop concerts. As a result of this and other incidents the report *Safety of Sports Grounds* was produced.

Ramsgate Walkway, 1994

The collapse of a steel framed walkway (see Fig. 5.6) to Ro-Ro ferry resulted in the death of six passengers and serious injuries to seven others. As a result of this, and after considerable discussion, a new Ro-Ro standard (BS 6349-8 *Code of practice for the design of Ro-Ro ramps, link spans and walkways*) has been produced in an attempt to safeguard against any repetition of this tragedy.

Piper's Row Car Park, 1997

Collapse of flat slab reinforced concrete structure, possibly due to top mat steel reinforcement corrosion. Building designed to CP114:1957 and constructed using lift-slab technique in 1964. A 120-tonne section of the top floor collapsed, due to a punching shear failure which

Figure 5.6 The collapsed Ramsgate walkway (J.C. Chapman in Campbell 2001).

developed into a progressive collapse. The event highlighted the need for professional and regular inspections of structures of this type. This incident highlights the need for regular inspection of aging structures, something rarely achieved in the UK.

Avonmouth, 1999

Death of four bridge workers following falling temporary access trolley. Primary cause was the absence of end stops on the temporarily open ended track sections which supported the gantry trolley. Partly as a result of this incident the IStructE report on access gantries has been updated to include temporary gantries.

Heathrow Tunnel Collapse, 2000

Collapse occurred during construction of a tunnel to be used for the Heathrow Express Rail Link. The site was between the airport's two main runways. The judge hearing the litigation case described the event as 'one of the worst civil engineering disasters in the last quarter of a century'. Main contractor fined £1.2m plus costs. The event illustrated the need for good communication between the parties concerned when using the observational method of tunnelling.

Gerrards Cross Tunnel Collapse, 2005

Collapse of 30m length of tunnel being constructed using precast concrete units above Chiltern railway line. Although not fully reported at the time of writing, this incident highlighted the need for robust designs and the careful monitoring of soil backfilling and other temporary works.

Buncefield Depot Explosion, 2005

Explosion at an oil storage facility in Hertfordshire described at the time as the biggest incident of its kind in Europe. Damage to adjacent property, including a school and a church together with domestic and commercial property, was widespread. The event brought into question the location of such a facility so close to other vulnerable property. Sound of the explosion was heard as far away as France and the Netherlands. The blast registered 2.4 on the Richter scale. Although not fully reported at the time of writing this incident highlighted the need for good maintenance of mechanical equipment and constant monitoring of the workings of process plants; it also raised questions about the siting of such facilities so near to domestic properties. Had this tragedy occurred mid-week the loss of life and injury might have been very high.

5.1.2 Learning from the past

The situation may have improved slightly since the 1990s but one is still aware of masonry cladding without proper support and adequate jointing, movement joints in brick bridge parapets bridged by stone copings, car parks with slack falls unable to shed rainwater, housing developments built with inadequate site investigation, scaffold and tower crane collapses and many other shortcomings. And, perhaps even more importantly, a poor record in monitoring the performance of structures, poor maintenance and management systems for the use of facilities. There are still many lessons to be learnt from past failures.

In order to better inform the industry and to enable lessons to be learnt from failures and near failures, the Standing Committee on Structural Safety (SCOSS) and its companion Confidential Reporting on Structural Safety (CROSS) are collaborating.

In considering refurbishment and repairs practitioners should take advantage of:

- Known dead-weights of structures that may be less than the original design dead-load.
- Excessive superimposed loads in old regulations and/or original design.
- The use of fire engineering techniques that enable a more accurate assessment to be made of fire risks and means of escape than was possible at the time of the original design.

5.1.3 Relocating structures

In considering refurbishment schemes there should be more awareness of some spectacular examples of complete buildings that have been moved bodily from their existing location to another (see Figs. 5.7 and 5.8). This will almost certainly involve temporary internal bracing of the existing structure and the formulation of a temporary or permanent supporting grillage beneath the superstructure to facilitate sliding the whole structure to a new location.

Figure 5.7 Photograph of Belle Tout lighthouse, moved because of coastal erosion (Courtesy Abbey Pynford).

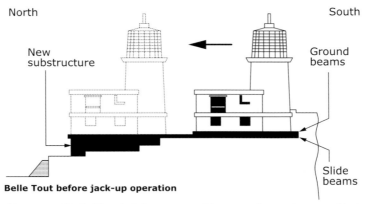

Figure 5.8 Diagram of Belle Tout lighthouse, moved because of coastal erosion (Redrawn after Abbey Pynford).

John Pryke (1987) postulates that the golden rules for these operations are:

- Never undermine load-bearing support.
- Never commence an operation that cannot stop indefinitely or be reversed.
- Aim to increase critical safety factors as the work proceeds.
- Keep things simple.

His company claims to have moved well over 60 buildings without serious problems. One of the heaviest structures moved by Pryke was a 3000 t bridge, which was slid into position in Wandsworth, London.

Figure 5.9 An illustration of transport for moving very heavy loads such as required in moving structures (Courtesy M. Wade, of Dorman Long Technology).

5.2 Basic materials: background, defects, strengthening and remedial measures

A useful general guide to the chronology of the use of materials is shown in Table 5.1 reproduced from the IStructE report on *Appraisal of existing structures.*

5.2.1 Concrete

5.2.1.1 General

Concrete is not a new material: it is claimed that the first example was its use in the floor of a hut in Jugoslavia in 5600 BC. Many examples from the Greek and Roman periods are documented. Until the late 19th century concrete was for the most part unreinforced, but at the end of the century under the influence of Coignet, Hennebique (a French contractor) and others the development of reinforced concrete (originally known as ferro-concrete) began. From early times it was recognised that concrete was strong in compression and weak in tension.

Table 5.1 Materials and/or form of construction and period of availability/use in the UK (based on Table 6.1 from *Appraisal of existing structures*, 2nd edn., IStructE). The density of the shading indicates the extent of the use.

Material and/or Form of Construction	Period of Availability/Major Use in UK (1600–2000)	Notes
Wood Based		
Timber frame with clay or brickwork infill: home-grown hardwood		
Timber frame with brickwork, or tile-cladding, softwood		
Timber frame and inner leaf with brickwork cladding		
Roof trusses and floors on masonry walls: home-gown hardwood		
Roof trusses and floors on masonry walls: softwood		
Bolted, and/or glued, laminated beams and frames		"Bolt-lam" used 1820–1920
Plywood – web and similar beams		
Natural Stone Masonry		
Loadbearing walls and piers: solid or with rubble cores		
Subsequent facing to timber frames and loadbearing brickwork		
Bonded facing to brickwork backing		
Cladding to steel, or reinforced concrete, construction		
Brickwork and Blockwork		
Loadbearing walls: clay bricks in lime mortar		
Loadbearing walls: clay bricks in cement* mortar		* Includes cement/lime mortars
Loadbearing walls: calcium–silicate bricks in cement* mortar		
Loadbearing walls: concrete blocks in cement* mortar		
Infill to timber frames: clay bricks in lime mortar		
Infill/cladding to steel, or concrete, frames: clay bricks, cement * mortar		
Infill/cladding to steel, or concrete, frames: calcium–silicate bricks		
Infill/cladding to steel, or concrete, frames: concrete blocks		

Period axis labels: 1600, 1700, 1750, 1775, 1800, 1825, 1850, 1875, 1900, 1920, 1940, 1950, 1960, 1970, 1980, 1990, 2000

Table 5.1 Materials and/or form of construction and period of availability/use in the UK (based on Table 6.1 from *Appraisal of existing structures*, 2nd edn., IStructE). The density of the shading indicates the extent of the use *(continued)*.

Material and/or Form of Construction	Period of Availability/Major Use in UK	Notes
Cast Iron		
Beams		
Roof truss components		
Columns		
Components for special structures: S.G. cast irons		
Water and gas mains, drain pipes, etc.		
Wrought Iron		
Tie rods, straps and bolts in timber structures		
Tie rods, chains, cramps and dowels in masonry		
Rolled strip, L and T sections		
Rolled I- and [sections		
Rivetted, built-up sections		
Wire for bridge suspensions cables (chains used earlier)		
Corrugated sheet (superseded by steel from 1890–1900)		
Structural Steel		
Plates and tie rods		
Rolled L-, T-, I- and [- sections (Is up to 610 × 190)		
Rivetted, built-up sections		
Structural tubes		
Rolled "parallel-flange" I sections up to 1100 mm high		
Rectangular hollow sections		
Welded, built-up sections		

Table 5.1 Materials and/or form of construction and period of availability/use in the UK (based on Table 6.1 from *Appraisal of existing structures*, 2nd edn., IStructE). The density of the shading indicates the extent of the use *(continued)*.

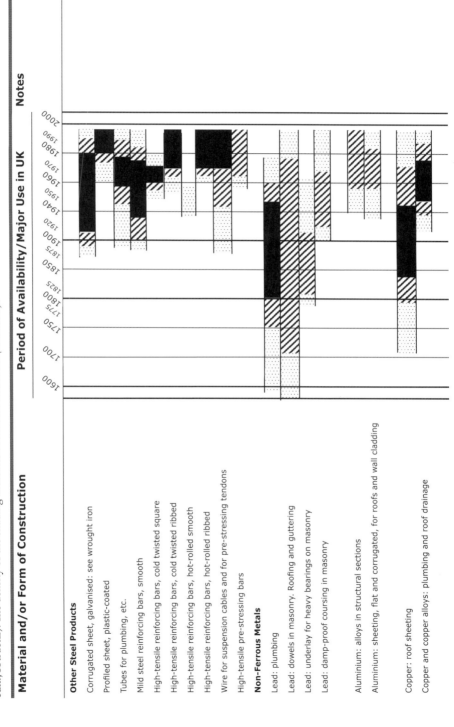

Table 5.1 Materials and/or form of construction and period of availability/use in the UK (based on Table 6.1 from *Appraisal of existing structures*, 2nd edn., IStructE). The density of the shading indicates the extent of the use *(continued)*.

Material and/or Form of Construction	Period of Availability/Major Use in UK	Notes
Cement and Concrete Based		
Mass concrete: lime or pre-Portland cement based		
Mass concrete: Portland cement based		
Clinker aggregate concrete in filler-joist floors		
Reinforced concrete: early patent systems		
Reinforced concrete: conventional design		
Pre-tensioned concrete		
Post-tensioned concrete		
Lightweight structural concrete, blocks and screeds		
Asbestos-fibre-reinforced cement sheet materials		
Glass-fibre-reinforced cement (cladding and permanent formworks)		
Other fibre-reinforced concretes (screeds and ground floors)		
Polymer-fibre-reinforced concrete		
Organic Polymer-Based		
Glass-fibre-reinforced plastic		
Polymer sheeting		
Miscellaneous		
Rammed earth "Cob"		Still in sporadic use
Wattle and daub		
Asphalt roofing, guttering and waterproofing		
Bituminous felt roofing		

(Period axis: 1600, 1700, 1750, 1775, 1800, 1825, 1850, 1875, 1900, 1920, 1940, 1950, 1960, 1970, 1980, 1990, 2000)

Figure 5.10 Weaver's Mill, Swansea (Courtesy Robin Whittle).

The late 19th century saw the development of several industrialised methods for producing structural frames for buildings. In the early 20th century L G Mouchel built on the experience of Hennebique to set up a consultancy which still exists today. Hennebique's system used plain round reinforcing bars with fishtail ends. In a later development he used bent up bars to provide continuity and enhance shear capacity. The oldest *in-situ* RC building in the UK was thought to be Weaver's Mill in Swansea constructed in 1897 but now demolished.

During the Second World War considerable use was made of reinforced concrete for the construction of marine structures. The survival, to this day, of the sea forts in the North Sea and remains of units of Mulberry harbours bear testimony to the durability of the material if properly specified, detailed and constructed.

The post war period (partly due to the shortage of steel) saw the rapid development of prestressed concrete. One fine example of this technique was the spectacular BEA hangar at Heathrow Airport. 'Laingspan' and 'Intergrid' were developed for school construction. Several precast flooring systems (eg Pierhead, Stahlton and Millbank) were developed; Tarmac used prestressed concrete methods for the manufacture of railway sleepers.

Construction increased considerably during the 1960s and in the early 1970s UK cement usage peaked at 18.5 million tonnes per annum. Also about this time the North Sea oilfields were being developed and the Condeep offshore platforms were constructed. However concrete found it difficult to compete with steel in these situations.

Dr George Somerville (1986), in an award-winning paper to the Institution of Structural Engineers, defined the essential requirements for the making of durable concrete by the four Cs – Constituents, Cover, Compaction and Curing. If insufficient attention was given to any of these elements then the likelihood of producing a good result would be slight (see Figs. 5.11 and 5.12). As previously

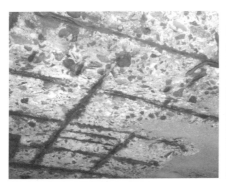

Figure 5.11 Result of badly positioned reinforcement (Courtesy Chris Shaw).

Figure 5.12 Properly fixed reinforcement: note spacers to achieve correct cover (Courtesy Chris Shaw).

reported, repairs due to poor attention to cover is costing the industry a staggering £550M per annum. Chris Shaw has masterminded the production of BS 7973 (a follow-up to the 2001 CIRIA report – *Specifying, detailing and achieving cover to reinforcement*). He has confidently stated that reinforced concrete that is detailed and constructed to comply with BS 7973 will always achieve the correct cover.

Well made concrete is very durable and has inherent resistance to fire (see Section 5.3.6.3).

In its early manifestations concrete was a relatively simple material. Before and immediately after the Second World War a cubic yard of structural concrete might contain Portland cement, natural coarse and fine aggregates and sufficient water for hydration; reinforcement would probably be plain round mild steel bars. The concrete would most probably have been volume batched. Compaction might have been achieved by hand-tamping or early types of surface or poker vibrators. Curing, if carried out, would have been by damp sand or hessian. 1:2:4 mixes achieving an allowable strength in compression of 3000 psi were the order of the day.

Today a cubic metre of concrete might contain a wide variety of cements, of which there are now 90 types manufactured by BCA companies. Cements may be blended with pulverised fuel ash (pfa) and/or ground granulated blast-furnace slag (ggbs) and/or microsilica. There should be sufficient water for hydration. The reinforcement might be a mix of mild, high tensile or stainless steel with plain or deformed cross-sections. The reinforcement could also be epoxy coated or galvanized. Various additives or admixtures to improve workability and/or to accelerate/retard strength gain might be present and the mix might be air-entrained (too much air-entrainment may lower the strength of the concrete). Most mix materials would be weigh-batched. It is also likely that concrete would be delivered to site ready-mixed and possibly pumped into position. Compaction would be achieved by sophisticated vibration techniques. Curing would most likely be by use of a sprayed chemical membrane. Characteristic strengths of concrete can vary between

2 N mm² (no-fines concrete) to 100+ N mm² (very high strength concrete). It may also be self-compacting. Whilst greatly enhanced properties can be achieved with modern mixes (up to and beyond 100 N mm²), the increased complexity brings with it more chance of error and loss of long-term durability.

Such a comparison highlights the need for a very careful examination of concrete before proceeding with repairs or modifications. In some cases it may be possible to strip the concrete back to the reinforcement, add additional bars and then re-face using sprayed concrete techniques. More advanced methods include desalination and/or cathodic protection to inhibit the growth of chloride induced corrosion (see Bibliography and further references).

In a world conscious of the need to reduce emissions of carbon dioxide, the Cement and Concrete Industry is claiming that 20% of the direct CO_2 emissions from the manufacture of cement are re-absorbed by concrete over its lifecycle.

5.2.1.2 *Prestressed concrete*

The primary objective in pre-stressing has been described by the late Dr SCC Bate as a technique 'used to avoid excessive cracking and deflection whilst at the same time enabling high-strength materials, particularly high-tensile steel to be used efficiently in construction.'

Pre-stress may be induced into concrete by stressing high tensile wires, rods or cables. Although pre-stressing has a good record, problems have occurred in prestressed precast units made on the long line system and used mainly in flooring, also in bridge designs using post-tensioning systems. Problems in flooring units were mainly due to the misuse of high alumina cement (HAC) causing rapid loss of strength due to conversion (see section dealing with HAC). Some examples of this are still current and some floors have been replaced or strengthened. Those floors where the HAC units act compositely with a structural concrete topping are less vulnerable to malfunction although engineers will usually recommend regular monitoring to check rates of deterioration (see also *High Alumina Cement*).

The trouble with post-tensioning systems was mainly related to corrosion of pre-stressing tendons due to faulty grouting of cable ducts. So severe were some of these problems that the DoT/HA placed an embargo on the use of post-tensioned bridges in their care from September 1992 to August 1996. Where inadequate grouting has been found, it has been necessary to re-grout or strengthen the bridge. The Concrete Society Report No 47, *Durable post-tensioned concrete bridges* (CS 2002), provides practice guidance for future work.

Non-destructive testing (NDT) techniques can effectively be used in detecting the extent of defective grouting. The extent of repair required for defective construction will depend on the extent of grout failure. In some cases it may be possible to re-grout the structure. In more severe cases, augmentation or replacement of the pre-stressing system will be required.

5.2.1.3 Glass-fibre reinforced cement (GRC)

GRC is a glass-fibre reinforced cement that is a combination of alkali-resistant glass fibres and a cement/sand mortar. The resultant material is similar to a fibre-reinforced concrete. Early experiments with this composite failed due to the vulnerability of the fine glass fibres to the high alkalinity of the cement matrix. Following collaboration between BRE and Pilkington, 'Cemfil' glass fibre was produced and proved to be more alkaline-resistant.

Since about 1975 under the guidance of the Glassfibre Reinforced Concrete Association (GRCA), the material has developed and is now used in a wide range of construction related products. Typical section thicknesses are 6–20 mm and are an alternative to precast concrete, sheet metal, timber, plastic and asbestos cement. Manufacturers claim advantages compared to other materials are less combustibility, better fire resistance and lower dead weight. Applications include cladding, permanent formwork, pipe-work and drainage channels.

Some examples of GRC cladding have exhibited surface crazing. Low dosage and/or poor distribution of fibres are likely to produce poor performance of units and may require early replacement.

5.2.1.4 Sprayed concrete

Sprayed concrete (sometimes known as gunite or shotcrete) is used for immediate temporary support following excavation; for repair or specialist purposes such as the construction of folded plate structures. When used for repair to damaged concrete it is essential that the damaged material is cleaned away and if necessary the reinforcement augmented to provide additional strength. It should be noted that, in spite of sub-contractor assurances, it will be virtually impossible to match the colour and texture of the existing concrete so a suitable coating will need to be considered if architectural appearance is important. Where a significant thickness of sprayed concrete is required best practice suggests that the material should be applied in layers so that the new material successfully adheres to the substrate. A mild steel or stainless steel mesh is often incorporated for additional strength.

In the New Austrian Tunnelling Method (NATM) observational method, sprayed concrete is used to form the tunnel lining (see also section dealing with tunnels). It should be noted that the use of accelerators and/or other additives in the sprayed concrete may adversely affect the final strength of the material.

5.2.1.5 Reinforced autoclaved aerated concrete (RAAC)

It is understood that a Swedish architect, Johan Ericsson, originally developed this form of construction in his quest to produce a lightweight building material with good thermal insulation and fire resistance.

RAAC is a material that differs in nature from normal dense concrete because of its different characteristics and material properties. Matthews *et al.* (2002) have suggested that the term concrete is inappropriate and that a more accurate description might be aerated (or cellular) silicate.

This material has been manufactured into precast panels to form wall and roofing units. It has also been used in unreinforced block form in beam and block floors. Usage was prevalent in the 1960s and 1970s: the most common units were marketed under the trade names DUROX© and SIPOREX©. This lightweight material generally falls into the range 500–700 kg/m³ (bulk dry density), with a compressive strength in the range 3–6 N/mm². The short term elastic modulus is in the order of 2 kN/mm².

Apart from the use in standard blockwork, units are typically reinforced using steel as used in normal reinforced concrete. Autoclaved aerated concrete (AAC) carbonates rapidly and therefore does not provide adequate protection to the steel normally afforded by the alkaline internal environment and its use has now effectively ceased. It is therefore necessary to protect the steel using either bituminous or cement-rich latex coatings, thus reducing the concrete bond. In some cases galvanized reinforcement is used.

Defects noted by BRE and others include:

- Corrosion of reinforcement
- Rapid carbonation
- Excessive deflections – leading to ponding of rainwater on flat roofs (see also section on *Flat Roofs*)
- Roof and floor units acting independently thus lacking the load-sharing ability of more conventional systems

Any plan to retain such structures in a refurbishment scheme should be supported by a thorough survey and test programme to assure long term adequacy.

5.2.1.6 Mundic

Mundic is the Cornish word for iron pyrites or fool's gold. Concrete blocks (and also mass concrete) made using aggregates derived from mine waste have been found to decompose badly. It is thought that sulfides within the material react with water and oxygen to form sulfate ions which attack the cement binder causing expansion, deterioration and loss of strength of the concrete typically evidenced as pop-outs in overlying plaster and cracking of wall sections. Early cases of this problem were detected in houses in the Perranporth and St Agnes areas but the problem is now widespread throughout Cornwall and South Devon.

It is difficult to suggest a solution to this problem other than, in severe cases, to rebuild parts of the structure. Any remaining sections containing Mundic must be fully protected against water penetration.

Table 5.2 Defects in plasters and renders.

Defect	Likely Cause	Remedial/ Preventative Action
Blistering of plaster	• Intense local movement of the final coat relative to the undercoat. • Delayed expansion of Class C and D plasters.] • Frost damage	• Ensure final coat is well keyed or bonded to undercoat and, for instance, don't expose plastering to local source of radiant heat when drying out. • Don't allow these plasters to dry too quickly after they have been applied or expose them to persistent moisture.
Bond failure - bossing and/or spalling of plaster or external render	• Weak background or undercoat. • Dense smooth background or undercoat. • Excessive thickness of plaster or render. • On expanded metal lathing use of wrong type of second lightweight undercoat. • Corrosion of metal lathing if stainless steel or galvanised material not used • Shrinkage of background and/or undercoat(s). • Insufficient wetting down • Water penetration through cracks in render and frost action • Misuse of bonding agent	
Cracking	• Movement in the background or surrounding structure. • Sulfate attack in the render	
Crazing	• Loamy sand • Excess lime putty in final coat • Applying final coat before initial shrinkage of undercoats is complete (if undercoats are based on lime or cement)	
Efflorescence	Excess salts in materials and/or substrate	Clean off salts
Flaking or peeling of final coat		
Grinning or irregularity of surface texture		
Popping or blowing of plaster	Particles in the background or in the plaster, lime or sand, which expand after the coat has set.	Remove pop-outs and infill or replace section of wall if cracking is too severe
Recurrent surface dampness on plaster	Deliquescent salts attracting moisture from the air. The salts often come from the use of unwashed sea sand. They can be carried from the background into the plaster by, say, condensation in an unlined flue.	May be possible to remove salts by repeated washing and then to stabilise surface
Rust staining	Application of inappropriate plaster to, say, unsuitable metal lathing or plaster in contact with corrodible ferrous metal in persistently damp conditions, e.g., where deliquescent salts are present.	Remove and replace in severe cases
Softness or chalkiness of plaster	Excessive suction of the undercoat, undue thinness of the final coat, working past the setting point, or exposure of final coat to excessive heat or draughts during setting.	Use stabilising solution in minor cases; otherwise replace

5.2.1.7 Renders and plasters

Renders are usually a cement-based coating applied to the exterior of buildings to prevent moisture penetration or internally to produce a uniform surface to which finishes may be applied. Cements used for brickwork mortars are usually suitable for renders. Additives to increase workability and water retention may be used to increase the effectiveness of renders. Shrinkage can be reduced by the introduction of fibres to the mix. The use of pigments can add to the aesthetic appearance of the material.

Renders should be applied in two or three coats with adequate curing time between each application. Backing coats should be 8–16 mm thick; second coats should be perhaps 8–13 mm and third coats 10–13 mm. It is also important that overlying coats reduce in strength outwards.

Defects encountered include:

- Sulfate attack derived from salts in wet brickwork indicated by yellow/white discolouration and cracking of the render in sympathy with brick jointing.
- Loss of adhesion due to poor workmanship, inadequate specification or frost action.
- Shrinkage possibly due to differential settlement or to too strong a mix.
- Breakdown of externally applied coatings such as cement paints.
- Too great a difference of strength between finishing and backing coats.
- Moisture penetration at edges of areas of rendering.

In most cases the damaged areas must be removed and replaced.

Historically, plasters have consisted of lime and sand, with either cement or gypsum added as a binder. However since about 1945 gypsum plasters (generally pre-mixed and only requiring the addition of water) have dominated the scene. There is a type of plaster to meet almost every requirement. These include:

- Carlite pre-mixed
- Thistle board finish
- Thistle one-coat
- Thistle renovating
- Thistle projection
- Thistle pre-mixed gypsum X-ray
- Thistle Hardwall
- Thistle multi-finish

5.2.1.8 Large panel systems

See Section 5.3.26 on cladding.

5.2.1.9 High alumina cement (HAC)

On 13 June 1973 the roof of the assembly hall at the Camden (London) School for Girls collapsed. On 8 February 1974 a similar collapse occurred to the roof over the swimming pool at Sir John Cass's Foundation and Redcoat Church of England Secondary School in Stepney, East London. In the case of the Cass School, failure occurred 18 years after construction. Extensive investigations into these incidents revealed that the use of high alumina cement in the precast prestressed concrete beams to these roof structures was the principal cause of collapse. Concrete made using HAC was of considerable advantage to precast concrete manufacturers because it gained high early strength thus enabling formwork to be struck early and immediately re-used. However, in certain conditions, the HAC concrete suffers from a phenomenon called conversion, resulting in a change in the structure of the hydrated cement. Under certain conditions this may be accompanied by a reduction in strength of the concrete. The degree of conversion is, however, complex and depends on the quality of the concrete and the temperatures to which it is subjected to throughout its life. Chemical attack and the presence of moisture may also be factors in the degree of conversion and subsequent loss of strength.

In severe cases of deterioration it will be necessary to replace parts or all of the structure. In less severe cases, especially where beams made from HAC concrete act compositely with *in-situ* Ordinary Portland Cement (OPC) concrete, the structure may continue to perform satisfactorily but need regular monitoring to check for any further deterioration. Recent investigations by BRE have revealed that HAC concrete is often carbonated to the depth of the reinforcement, therefore heralding the onset of reinforcement corrosion. The use of HAC in structural concrete is now effectively banned but it may still be used for refractory purposes. It should also be noted that one of the effects of using blends of HAC and OPC may be to induce a flash set.

5.2.1.10 Alkali-silica reaction (ASR)

ASR is one example of a more general problem known as Alkali Aggregate Reaction (AAR). ASR is a chemical process in which alkalis, usually predominantly from the cement or external contaminants (e.g. de-icing salts) combine with certain types of silica in the aggregate when moisture is present. This reaction produces an alkali-silica gel that can absorb water and expand to cause cracking and disruption of the concrete (see Fig. 5.13). For a damaging reaction to take place the following need to be present:

- High alkali cement
- Reactive aggregate (e.g. crushed greywacke type sandstone, microcrystalline quartz or chalcedony found in flints and cherts)
- Moisture

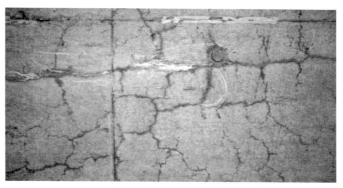

Figure 5.13 Cracking of reinforced concrete due to ASR (Courtesy Dr Jonathan Wood).

An interesting phenomenon concerning ASR is the 'pessimum'. This is defined as the reactive constituent in the total aggregate which yields the greatest resultant concrete expansion. Curiously, if the content of reactive material is either reduced or increased from the pessimum the amount of expansion will decrease.

Structures most at risk are bridges, hydraulic structures, exposed frames (e.g. open multi-storey car parks) and foundations. In relation to total construction volume, the incidence of significant damage due to ASR in the UK is small. ASR was discovered in the USA in the 1940s; no case was found in the UK until 1971 (in a concrete dam in Jersey) although recent publicity has revealed that the Montrose Bridge in Scotland (now demolished because of ASR) was about 75 years old.

The problem is known to exist in at least 46 countries (see Table 5.3). Most have specifications in place to minimise the likelihood of occurrence. Measures include limiting the alkali in the cement, use of non-reactive aggregates, shielding the structure from moisture and the use of blends of cements by introducing various dosages of pfa, ggbs and silica fume. Most of these recommendations approximate to UK practice.

Affected concrete often exhibits map cracking of the surface, known colloquially as Isle of Man cracking after that island's three legged emblem. In heavily loaded sections the cracking will probably follow the line of the main reinforcement.

Guidance on the identification is given in BCA report *The diagnosis of alkali–silica reaction* however if ASR is suspected it is essential for clients to obtain a report on the condition by a materials test house that has the benefit of a structural engineer. Mild cases of ASR may be managed by using a combination of external coatings; restriction of the supply of moisture and regular monitoring of the structure. More severe cases may require major surgery and member replacement.

The blending of cement with pulverised fuel ash (pfa) and/or ground granulated blastfurnace slag (ggbs) in correct proportions has proved effective in minimising ASR. There is also some evidence that the inclusion in the concrete mix of metakaolin (a by-product of china clay) has a suppressing effect on reactive aggregate.

Table 5.3 ASR throughout the world (Courtesy Dr Darrell Leek, Mott Macdonald).

	Europe		Americas		Africa		Asia
1	Austria	27	Argentina	31	Algeria	38	Australia
2	Belgium	28	Brazil	32	Egypt	39	China**
3	Bosnia Herzegovina	29	Canada	33	Ghana	40	Japan
4	Croatia	30	USA	34	Kenya	41	India
5	Cyprus			35	Mozambique	42	Iran
6	Denmark			36	South Africa	43	Iraq
7	Eire			37	Zimbabwe	44	New Zealand
8	Finland					45	Pakistan
9	France					46	Taiwan
10	Germany						
11	Greece						
12	Iceland						
13	Italy						
14	Macedonia						
15	Netherlands						
16	Norway						
17	Poland						
18	Portugal						
19	Romania						
20	Serbia						
21	Slovenia						
22	Spain						
23	Sweden						
24	Switzerland						
25	Turkey						
26	United Kingdom*						
*	Includes cases in England, N. Ireland, Scotland, Wales and the Channel Islands						**Includes cases in Hong Kong

5.2.1.11 Deleterious aggregates

It should not automatically be assumed that aggregates will produce satisfactory concrete (see Figs 5.14 and 5.15). Potential problems include:

- High concentrations of chlorides in aggregates from marine and other sources. Design codes limit the level of these to safe levels. For example

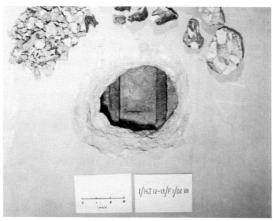

Figure 5.14 Cavitation in 5" thick concrete slab due to high concentrations of chlorides in aggregate.

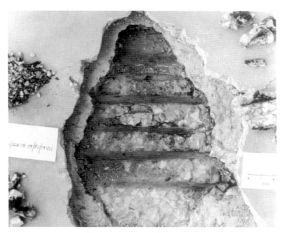

Figure 5.15 Deterioration in reinforcement in 5" thick concrete slab due to high concentrations of chlorides.

BS 8110 suggests upper limits (expressed as a % of chloride ion by mass of cement) of 0.1, 0.2 or 0.3 respectively for prestressed concrete, concrete with BS 4027 or BS 4248 cement or concrete containing embedded metal and manufactured with various cements or in combination with ggbfs or pfa. Recently introduced Eurocodes require similar limits. Historically, many reinforced concrete structures built in the Middle East around the 1960s were found to be defective and were subsequently demolished. After much research it was found that the very high chloride content in the aggregates was the cause of an electrochemical reaction producing rapid corrosion of the steel reinforcement. Much of this aggregate had been obtained from highly contaminated desert surface deposits. Many of these problems centred on Bahrain. Once the problem was identified cleaner aggregates were imported from Ra's al Khaymah and elsewhere.

- Sulfates are present in some aggregates and can cause expansion and disruption to the concrete. The total water soluble content of the concrete, as expressed as SO_3 should be limited to 2.5–5% by mass of the cement in the mix. Again, Eurocodes express similar caution.
- A number of naturally occurring aggregates (e.g. greywacke) are considered to be of high reactivity and may, combined with high alkali cements in the presence of water induce alkali-silica reaction (see Section 5.2.1.10). Furthermore in the absence of additional research it is suggested that all recycled aggregates be considered reactive. The actual amount will be dependent on the type of cement.
- Some years ago some Scottish aggregates were deemed highly shrinkable and thus not suitable for the manufacture of satisfactory concrete.
- The 16th Report of SCOSS highlights research that for reinforced concrete members utilising limestone aggregates and when unreinforced in shear, the resistance may be below the characteristic strength anticipated by the codes. This is specifically the case for high strength concretes.

5.2.1.12 Carbonation

When carbon dioxide in the air combines with rainwater it forms carbonic acid. The alkalinity of the protective concrete cover to reinforcing steel is reduced by the carbonic acid so that water and oxygen attack and corrode the steel. This neutralisation is known as 'carbonation'. The rate at which it proceeds from the surface depends on a number of factors such as porosity, type of cement etc. One authority has quoted that carbonation proceeds at a rate of 5–10 mm every 10 years. The degree of carbonation can be established by sampling and testing.

In extreme cases it may be necessary to remove the existing cover, clean and/or augment the reinforcement and rebuild the cover using spray concrete techniques and, if necessary, other repair methods to provide adequate protection against further carbonation. It should be noted that sprayed concrete often carbonates rapidly.

5.2.1.13 Hydrogen embrittlement

This comparatively rare phenomenon is not fully understood. It is predominantly applicable to small-diameter highly stressed steel used as reinforcement in prestressed concrete and may lead to brittle-like fracture of the material. It is caused by a penetration of the metal lattice by atomic hydrogen and may be drawn to areas of high stress and surface defects in the steel. It has been stated that if hydrogen is already present in the metal lattice, before the application of load, then a delayed fracture may occur.

It can also occur in high tensile bolts. Eminent bridge engineer Charles Cocksedge has put this succinctly by stating: 'It's the reaction between a high tensile steel and water. If there's a metal coating as well, the two metals react and water liberates hydrogen which embrittles the steel and can cause premature failure'.

Replacement or strengthening of the structure will be necessary if hydrogen embrittlement has taken place.

5.2.1.14 Rust staining

An appearance of rust on the surface of concrete may be evidence of real or superficial damage to reinforced concrete structures. This may be caused by one of the following:

- The presence of corroding reinforcement near the surface. This may be the outward manifestation of serious corrosion of reinforcement or simply corroding tie wire. The rusting may also be accompanied by cracking of the concrete in the area of rusting In this case a thorough investigation must be carried out and may result in removing the damaged concrete and repairing using sprayed concrete techniques or even removing part of the structure and rebuilding the damaged area.
- Alternatively the apparent rust staining may be due to the presence near the surface of ferrous sulphide (iron pyrites) inclusions in which case the problem may be classified as unsightly but not unsafe. If considered too unsightly then the provision of a coating or other protective treatment should be considered.

5.2.1.15 Acid and sulfate attack (including thaumasite)

In general, acids are damaging to concrete. Damage is typically caused by sulfuric and acetic acids and ammonia substances. These dissolve the more soluble constituents of the set cement leaving only a mushy deposit. Other substances such as milk also attack concrete. Deterioration is directly related to permeability of the concrete – the more permeable the concrete, the more deterioration.

Concrete used in sewer construction may be attacked by sulfuric acid produced by microorganisms from the waste material. Sea water may contain high concentrations of carbon dioxide and damage submerged structures such as slipways.

Many acids used in commercial premises may attack concrete. Concrete surfaces may be treated with an acid resistant coating. In cases where concrete deterioration has taken place it may still be possible to prolong the life of the structure by applying a coating. Where the deterioration is moderate it may be possible to grit blast the surface and then replace the cover by spray concrete or other hand-placed methods and then coat the surface. In extreme cases, the structure should be demolished and replaced.

Sulfate attack on cement-based materials has been recognised for many years and is normally minimised in the UK by the use where necessary of sulfate-resisting cement (SRC). In 1999 an Expert Group chaired by Professor Leslie Clark reported on a new form of foundation sulfate attack now termed thaumasite sulfate attack (TSA). The drivers for this type of attack were identified as:

- Presence of sulfates and/or sulfides in the ground.
- Presence of mobile ground water.
- Presence of carbonate, generally in concrete aggregates.
- Low temperatures (generally below 15°C).

Secondary influential effects include type of cement; quality of concrete; changes to ground chemistry resulting from construction and type; depth and geometry of buried concrete (see also Section 5.2.1.20). Remedial strategies depend on severity of attack but include monitoring, measures to avoid contact with water and, in most severe cases, complete replacement.

5.2.1.16 Woodwool formwork

In the early 1970s channel reinforced woodwool slabs were sometimes used to provide permanent decking formwork to reinforced concrete structures. In a number of cases it was discovered that spacers to reinforcement had penetrated the woodwool leaving little or no soffit cover to the reinforcement. In an extreme case (witnessed by the author) when the woodwool soffit was removed the bottom reinforcement to a beam fell to the floor below. Investigations into these failures also discovered bleeding from wet concrete into the woodwool leaving a layer of no-fines concrete in the structural components.

In 1975 at New Malden House some floor sections cast on woodwool formers failed as a result of using this type of permanent formwork.

In these cases it will be necessary to replace the affected structure; in less severe cases it should be possible to strip out the woodwool and replace the concrete cover using sprayed concrete or other techniques.

As a result of these problems, it is possible that some structures may exist where slurry- or mortar-coated woodwool slabs have been used for shuttering.

5.2.1.17 Repair, strengthening and remedial methods

Structural concrete may be repaired and strengthened in a variety of ways. These include:

- using sprayed concrete methods, possibly incorporating mesh reinforcement, in which existing sections are trimmed back to the main reinforcement etc
- adding steel sections to the existing structure
- using carbon fibre bonded externally
- using hand-applied cementitious repairs
- re-shuttering and using flowable concrete to re-cast to original profiles (see Fig. 5.16)

5.2.1.18 Cathodic protection

Cathodic protection has been described as a process of protecting a metal component

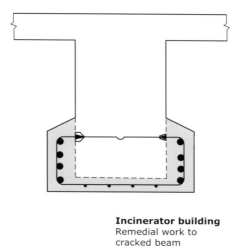

Incinerator building
Remedial work to
cracked beam

Figure 5.16 Strengthening of concrete beam.

or structure from corrosion by the installation of a sacrificial anode or impressed current system that makes the protected object a cathode and thus resistant to corrosion. The system is thought to have had its origins in work by Sir Humphrey Davy in 1824 who was investigating the corrosion of copper bottomed ship hulls.

These methods may be used retrospectively on damaged structures or applied to structures at the initial construction stage. The costs of these techniques are usually high and also need to be continuously monitored to check continuing efficiency.

5.2.1.19 Desalination and re-alkalisation

Desalination and re-alkalisation are similar corrosion-inhibiting methods in application but different in approach. Desalinisation effectively reduces the level of chlorides in the concrete although not necessarily from behind the steel reinforcement whilst re-alkalisation increases the alkalinity of the concrete by generating new hydroxyl ions. For further information see CPA State of the art report and other relevant literature.

5.2.1.20 Coatings

Coatings for concrete may be used, either initially or retrospectively, for a number of reasons including;

- Protection from atmospheric exposure
- Architectural appearance
- Protection against aggressive gases, liquids or other substances in container vessels

- To reduce the permeability of the base concrete.
- To improve waterproofing characteristics

There are many types of coatings which include:

- Polymers (acrylic and styrene/acrylic)
- Cement-based and fibre reinforced
- Epoxy resin
- Polyeurethanes
- Silanes, siloxanes and silicone resins
- Silicate paints
- Lime wash
- Anti-carbonation coatings
- Bituminous products

Biczok (1972) deals comprehensively with protection of concrete against attack by, for example, soft water, peat and marsh soils, alcohols, acid gases, carbonic acid, magnesium salts, chlorine, ammonia gas, sulfates and sewage.

5.2.1.21 Industrialised building systems

Under government and other pressures there have been many attempts to speed up the building process by the use of industrialised building systems. Many of these have involved off-site fabrication of units that are then brought to site and assembled. In the 1960s many systems were introduced, few of which are currently available. Many of these systems were related to the housing market, but other systems did exist for the construction of schools and other buildings. These systems include:

- Housing
  - *In-situ* concrete system BRE Report numbers are indicated thus: (BR191).
  - Wimpey no-fines (BR191 and BR153). No-fines concrete used in a load bearing mode for buildings (mainly domestic dwellings but also university hostels) up to five storeys; in conjunction with a reinforced concrete frame up to 25 storeys in high-rise flats. Approximately 750 high-rise blocks were built using this system.
  - Although Wimpey was the major sponsor of this system some dwellings were built by The Scottish Special Housing Association (SSHA) and Irish Contractor Farran.
- Other concrete systems
  - Mowlem (BR190)
  - Forrester-Marsh (BR155)
  - Cast rendered no-fines housing (BR156)
  - Incast housing (BR157)
  - Universal housing (BR158)

  - Fidler houses (BR 159)
  - BRE Type 4 houses (BR161)
  - Easiform cavity-walled dwellings (BR130)
  - Underdown and Winget houses (BR55)
  - Bison large panel system dwellings (BR118)
  - Lilleshall (BR469)

Precast concrete systems were also quite widely used in the 1950s to 1970s for schools, offices and other buildings. The components were mainly reinforced or post-tensioned construction for floor and roof structures. In particular, Intergrid and Laingspan were used for large span assembly halls in schools and similar structures (see SCOSS website). During the 1970s, evidence of corrosion of tendons in post-tensioned components, associated with exposure to moisture, were found in some of these buildings. In some instances this was due to the presence of chlorides. SCOSS have strongly recommended that these buildings are regularly inspected by experienced civil/structural engineers to check on potential problems.

5.2.2 Masonry

See also Section 5.3.6.5, 'Fire in masonry structures'.

5.2.2.1 General

Masonry usually comprises particulate materials such as brick and stone laid in a cement or lime based mortar. Stonework may be laid in a random or coursed manner. Many structures still in existence are constructed of load-bearing masonry.

Masonry includes brickwork, blockwork (ceramic, concrete etc), stonework, *in-situ* concrete (e.g. no-fines concrete and, occasionally, other materials). No-fines load-bearing construction has been widely used up to five storeys. Load-bearing brick has been used in the UK up to 14 storeys (e.g. Essex University). It is understood that higher load-bearing structures have been built in Switzerland. The Essex residential towers were of black, graduated high-strength, wholly load-bearing brick, 9 in thick internally and 11 in thick externally. No doubt being high-density bricks these would have been constructed no more than six courses at a time. In solid load-bearing brickwork it is important in highly stressed work that the appropriate type of brick is used and that bricks are laid frog up with the frogs completely filled with mortar during construction.

Prior to the First World War most load-bearing construction was of solid brick or block; post-Second World War an increasing awareness of thermal insulation requirements and waterproofing has led to the widespread use of cavity construction, particularly for low-rise housing. A conventional cavity is normally 2 in (50 mm) wide. Cavity construction varies between brick and brick, brick and block and other combinations of materials. For adequate performance the external skin should be provided with weep-holes to allow penetrative water to escape. These weep-holes usually take the form of open perpends. More recently, plastic formers inserted into perpends, may be used.

Cavity brickwork can usually be recognised by the presence of weep-holes just above foundation level and above lintels and supporting beams; also if in unrendered brickwork by the presence of stretcher bond. If confirmation is required a survey using thermal imaging or a tie detector will pinpoint the presence of ties inserted between the skins of masonry. In well designed load-bearing cavity construction the two leaves should be tied together to allow for stability and a better distribution of load between the leaves. Cavity ties may be of flat metal (no longer permitted unprotected mild steel, galvanized steel or stainless steel) or twisted wire (stainless steel or copper). If these are defective through omission, corrosion or excessive spacing there are methods of replacement available which avoid the problem of re-building the wall.

Careless brick or block layers often allow mortar droppings and other debris to remain at the base of the cavity and on ties thus nullifying the thermal and water resistance of the wall. Such malpractice can be easily avoided (see *Site Engineers Manual* (Doran 2009) Chapters 10 and 16).

Perforated bricks or blocks may also be used in wall construction to reduce weight on foundations and to increase thermal resistance. Brickwork can also be reinforced using vertical steel bars through holes in the individual bricks or by the insertion of lateral steel along bedding joints.

Correct construction demands that load-bearing masonry is provided with an adequate damp proof course (dpc) placed at a level just above that of the surrounding ground. Dpcs may be of slate, plastics or other impervious material such as engineering brick, or, more recently but rarely used, bituminous felt.

In recent times, and with varying degrees of success, cavities have been filled with foam to improve thermal resistance. As an alternative, slabs of polystyrene or similar materials have been built into the cavity as work proceeded. Designers need to consider whether such techniques reduce the waterproofing characteristics of open cavity construction. Refurbishment may require remedial work (by chemical injection or other means) to prevent further damp penetration.

Practitioners should be aware of the need to stabilise new brickwork when bricking-up existing doorways or similar openings. This can be done by toothing in the new work or by using specially designed stainless steel assemblies which can be bolted to the exposed edges of existing work (see Fig. 5.17).

Stone may be used in load-bearing situations, as cladding or as a decorative material. Cathedrals (such as Lincoln) and other massive buildings were often constructed of material won from local quarries; however there were exceptions such as Canterbury and Chichester, cathedrals which used French limestone. Stones used for construction fall into three categories: igneous (e.g. granite), metamorphic (e.g. slate) or sedimentary (e.g. limestone or sandstone).

Figure 5.17 Stainless steel wall extension device to facilitate toothing of new masonry to existing work (Courtesy Maureen Doran).

Stone deteriorates due, principally, to two factors – the weathering agent (frost, soluble salts, acid deposition and moisture/temperature cycles) and the chemical/physical structure of the stone.

Refurbishment work may call for cleaning or repair of stone. Cleaning techniques must be related to the type of stone and the intensity of the deposited dirt. Received wisdom suggests the initial use of gentle techniques such as low pressure water and brush proceeding as necessary to high pressure water, grit blasting and even chemical cleaning. In cases of severe sulfate crusting it may be necessary to mechanically remove this before proceeding with cleaning. In all cases health and safety precautions (such as the COSHH regulations) must be strictly observed.

Where stonework has decayed the following sequence of events should be followed:

- de-scale to remove any loose material
- dress back the stonework to a sound face
- replace with new stone, or
- dress back the stonework and re-build in suitably specified mortar – so-called 'plastic repair'

Commercial considerations will usually dictate which technique to use. It is often less expensive to replace stone rather than re-build with mortar. Care must be taken not to bridge over original joints (including movement joints). If, by cutting back, the load-bearing characteristics of a structure are likely to be impaired, it is essential to consult a structural engineer to check if any temporary propping is necessary.

5.2.2.2 Tudor brickwork

This is a type of construction formerly used in industrial buildings where external brick panels were wedged in between the flanges of uncased structural steel columns to give lateral support against wind and other lateral forces. The external face was exposed to the atmosphere and corrosion of the steel columns was commonplace.

5.2.2.3 Terracotta and faience

Terracotta is a dense ceramic material formulated from once-fired clay. It is usually reddish in colour (although a buff coloured alternative is available) which is usually derived from the Etruria marl raw material. Terracotta is mainly used as a decorative feature to embellish the elevations of buildings where it is secured by sand/cement mortar or by cramping systems. An alternative use is in decorative chimney pots, finials and gargoyles. Cramping systems are prone to corrosion in which case areas may need stripping and replacement. This may cause supply chain problems and require special arrangements with manufacturers' to make limited replacement supplies to order.

A modern use of terracotta is as rain-screen cladding as introduced by Renzo Piano on the Potsdammer Platz in Berlin. In such cases the terracotta units are supported on an aluminium sub-frame.

Faience is terracotta that has been glazed prior to a second firing. It can be seen in decorative slabs around building entrances. Due to problems with shrinkage in manufacture, the size of units is normally limited to 450 × 300. The rear faces are usually profiled to improve adhesion to backing mortar although in taller buildings copper cramps may have been used to secure the units. Although the glazing produces an impervious surface problems may arise due to water penetration through the joints setting up freeze thaw conditions. Refurbishment may simply comprise re-fixing but in more serous cases units will need to be replaced. Replacement will require special orders to be placed with manufacturers often necessitating long lead times.

5.2.2.4 Defects, repair and strengthening

Masonry materials are susceptible to damage by a number of agencies, as summarized in Table 5.4.

5.2.2.5 Moulds lichens and other growths

Since the introduction of the Clean Air Act in 1956 the prevalence of lichens and other similar growths attaching to external masonry has increased. In some cases these growths may obscure carvings, trap water and cause damage due to freeze-thaw cycles. Acid compounds from lichen may also erode limestone substrates and mortars. Such attacks are rarely the cause of significant structural damage but may be the cause of dampness in a building and/or the gradual breakdown of the external façade.

Table 5.4 Typical masonry defects.

Defect	Likely cause	Remedial action
Cracking	• Hairline cracking – frost action; sulfate attack; thermal or moisture movement, particularly in calcium silicate (sand lime bricks). • Other cracking – thermal/ moisture induced movement; structural actions.	• Consider filling in cracks, if non-structural, with lime-based mortar. • For structural cracks the remedy depends on the cause. Subsidence, for example, may require underpinning.
Erosion	• Abrasion by wind driven dust and sand. • Acid rain pollution. • Bacterial contamination.	• Carefully scrape off loose and friable material. • Consider applying breathable clear sealer (Douglas 2006).
Bossing and/ or spalling of surfaces	• Salt crystallisation (NMAB 1982). • Frost action. • Repointing using cement mortar instead of lime mortar. • Sulfate attack.	• Minimise exposure to moisture if possible. • Carefully scrape off loose and friable material. • Repair/re-point with a lime based mortar.
Indentation and breakage	• Impact damage. • Vandalism.	Depnding on extent of damage: • Plastic repair – using appropriate mortar and fine stone aggregate. • Indentation repair with similar stone use.
Staining and soiling	• Algal growth. • Mould growth. • Dirt.	• Consider masonry cleaning system – preferably a dry, non-abrasive type. Apply biocide to surfaces affected with organic growths.
	• Efflorescence. • Ferrous metal fixings or impurities in the stone causing rust stains.	• Lightly scrape off with a stiff brush. Avoid wetting if possible as this will eventually trigger later outbreak of salt/rust staining.
	• Chemical staining – particularly coloured yellow, orange, etc.	• Apply chemical poultice to affected surfaces to remove or neutralise contamination.
	• Oil or other liquid splashes.	• Apply chemical neutralising solution to affected surfaces.

It is also worth noting that nitrifying bacteria may also attack stone, concrete and mortars. Where it is thought prudent to apply remedial measures thick surface growths should be removed with a stiff brush and the residue treated with biocide chemicals.

5.2.2.6 Masonry ties

The most prevalent types of cavity masonry are *brick and brick* or *brick and block*. In order to function as designed, it requires the two leaves to be adequately tied together. Galvanised steel ties have now given way to more durable stainless steel ties. For cavities up to 75 mm wire ties are common practice with rectangular cross section rigid ties being used for wider cavities.

The industry has a poor record in the placement of ties which are often missed out and/or poorly placed. A survey using an infra-red camera will pinpoint omissions and techniques are available to replace ties without extensive demolition (see BRE 1988 (Digest 329)).

5.2.2.7 Efflorescence

Efflorescence is the manifestation of a whitish powder on the face of clay brickwork. It is normally present at the time of construction. The whitish powder is normally soluble salts from the body of the brick. During the life of the building the brickwork is at its wettest when it is first laid. As it dries, by evaporation from the front face of the brick moisture migrates to the surface carrying salts in solution. As the moisture evaporates from the surface the salts crystallise and are left as a white bloom. Although temporarily unsightly these salts will usually be dissolved by falling rain and will disappear. Occasionally efflorescence may lead to decay of the brick if salts crystallise beneath the surface of the brick.

5.2.2.8 Brick chimneys

A chimney stack may be defective for a variety of reasons including:

- Sulfate attack on the mortar in unlined or pargetted stacks caused by corrosive flue gases.
- Deterioration of mortar joints causing the stack to tilt to a degree that causes instability.
- Brick erosion due to wind, rain and frost.
- Earthquake damage.

Binocular surveys can be useful in making initial inspections of exposed sections of flues. If further examination is required then the flue should be swept and cleaned and a further examination made using a fibrescope (or borescope) inserted into pre-drilled holes in mortar joints

Remedial work will need to consider the cause of deterioration. Consideration can be given to:

- A complete rebuild of the external projection of the stack.
- Lining the chimney with a suitably protected sectional metal liner.
- In cases where lower reaches of a chimney have been removed then the upper reach (remaining) should be supported on gallows brackets. It is important that a check is made of the supporting wall to ensure structural adequacy.
- If the external section of the chimney is part of a listed building or other similar edifice of architectural merit then it may need to be rebuilt in its original style regardless of any further use of the flue.
- Redundant stacks can be reduced to below roof level and capped off with stone or slate capstones.

- In cases where listed buildings are being refurbished it may be appropriate to use lime rather than cement mortar.

5.2.2.9 Brick matching and cleaning of brickwork

Where listed buildings are being refurbished it may be appropriate to use lime rather than cement mortar. In situations where bricks need to be matched for replacement, advice may be obtained from:

- The Building Centre
- H Butterfield Ltd. (Selbourne Road, Luton, LU4 8QF, tel: 01582 491100), who have a brick library with *c*.1400 samples covering all bricks manufactured in the UK together with a selection of foreign samples. They provide a brick matching service
- The Bulmer Brick and Tile Company
- Charles Brooking, an expert in these matters, may be contacted via Jenny Lynch, tel: 020 8331 9312

There may be particular problems in cleaning existing brickwork, particularly on listed buildings where high pressure water cleaning may damage the face of the brickwork. Some guidance on this subject is given in *BRE Digest 280*. For stability and other reasons new brickwork (or block-work) must be adequately bonded into existing work. Where this is impractical, proprietary, stainless steel wall extension profiles should be used (see Fig. 5.17).

5.2.3 Metals: background, defects, strengthening and remedial measures.

See also Section 5.3.64 'Fire in metal structures'.

5.2.3.1 General

Although a number of individual metals are included in this section it should be noted that severe corrosion may take place where two dissimilar metals are in contact. This is known as bi-metallic corrosion, the severity of which is dependent on a number of factors including the relative nobility of the two metals (the greater the difference in nobility the more severe the corrosion).

Cast iron, wrought iron and steel are dealt with below. It is beyond the scope of this book to deal with other metals but readers are referred to *The Construction Materials Reference Book* (Doran 1992) for further information on metals including aluminium, copper, lead, nickel, tin and zinc.

5.2.3.2 Cast iron

Cast iron (CI), as the name suggests, is formed into sections by pouring the molten metal into moulds, usually of sand or loam. It is composed of iron and small percentages of carbon (2–4.5% but generally in the range 2–4%). CI may be divided into three main categories:

- Historic CI (mainly of interest to structural engineers) which is generally grey cast iron manufactured between 1780 and 1880.
- Modern CI which is virtually the same as historic iron but is generally of a higher quality and conforms to British Standards. It is used mainly in mechanical rather than structural engineering
- Ductile CI or spheroidal graphite CI which, although seldom used in construction, may have a future as a structural material. It conforms to British Standards and is a relatively modern material manufactured from about 1946.

The manufacturing process often results in a pitted surface texture by which the material may be recognised. CI is weak in tension and strong in compression (one fifth to one sixth as strong in tension as compression). It is not uncommon to find structural sections with tension flanges larger than compression flanges. The sections tend to be geometrically rather coarse with internal corners rounded and external corners sharp. Flanges to beam sections are often fish-bellied (see Fig. 5.18). The casting process leaves something to be desired in terms of quality control, therefore the mechanical properties tend to be somewhat variable. However limiting values have been established and can be found in the bibliography. In cases of doubt it is prudent to conduct strength testing on samples extracted from the structure.

The first cast iron beams appeared around 1792; the use of cast iron for all practical purposes ceased around 1914 with the outbreak of the First World War. Many cast iron columns were of hollow circular configuration although recent analysis has shown these to be of often eccentric construction (see Fig. 5.19). Guidance on mechanical strengths is given in London County Council (LCC) publications and in authoritative books by E H Salmon and others. For background information

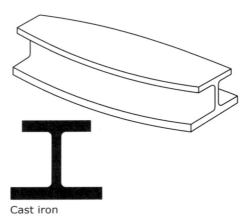

Cast iron

Figure 5.18 Cast iron beam: typical shape.

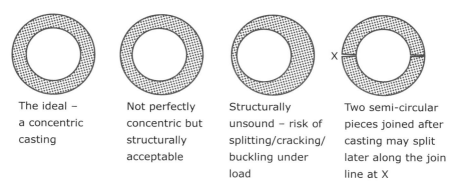

| The ideal – a concentric casting | Not perfectly concentric but structurally acceptable | Structurally unsound – risk of splitting/cracking/ buckling under load | Two semi-circular pieces joined after casting may split later along the join line at X |

Hollow circular cast iron columns are rarely found to be concentric and are often made up from two halves: problems can ensue.

Figure 5.19 Cast iron column sections indicating potential faults.

on the development of iron it is worth a visit to the website detailing the works of Henry Cort (see appendices for details).

One of the earliest classic structures in the UK was the Ironbridge in Coalbrookdale. Many distinguished buildings were also constructed using this material including the Crystal Palace building for the 1851 Exhibition.

Cast iron can be welded but this is best done in factory conditions. A recent example of this is the repair of the 1802 Iron Bridge at Stratfield Saye where large sections were removed, repaired and then replaced. *In-situ* repairs to cast iron can be carried out using a cold repair method of stitching such as that promoted by Metalok. Alternative methods of repair include stitching. This technique was used in the repair of Battersea Bridge, London the underside of which was damaged by collision with a vessel.

5.2.3.3 Wrought iron

Wrought iron is formed by passing billets of iron through rollers. The constituents of the material are iron and small quantities of carbon (0.02–0.05%). The appearance of the material is similar to steel. Early sections were built up by riveting together plates and angles. Later saw the appearance of rolled sections not dissimilar to modern day rolled steel sections. Generally sections were smooth-surfaced except where corrosion had occurred when lamination was present. The tensile strength of wrought iron was superior to that of cast iron and led to its extensive use in chains, cables, tie-rods and also links for suspension bridges. Composite sections made up of combinations of cast and wrought iron also appeared. A favoured building construction for many years was for cast iron columns used in conjunction with wrought iron beams.

The life span of the use of wrought iron for structural purposes was 1840–1914 when it was largely superseded by steel.

5.2.3.4 Steel

General

Structural steel is visually similar to uncorroded wrought iron. Some of the early, larger rolled sections were stamped with the makers name on the webs of beam sections. The principal ingredients of the material are iron, carbon and manganese. Carbon content is usually below 0.25% and manganese may vary between 0.5% and 2.5%. Proportions of these materials are varied to match the required strength, toughness, durability, weldability and appearance of the steel. Steel is easily recycled, although metallurgists warn of the possible presence of impurities (tramp metal) in the recycled product. CORUS (the steel industry trade association) claim that nearly 100% of the material is recycled.

In 1877 the Board of Trade (BoT) approved the use of steel for bridge building. The well known company Dorman Long rolled the first joist section in 1885 and details of these sections appeared in early section books. A great deal of information is to be found in the BCSA publication, *Historical Structural Steelwork Handbook* (Bates 1984).

Following the adoption of the convertor process in steel making (*c.*1856), the use of the material for structural purposes gathered pace. However it is understood that the first steel warship (HMS *Iris*) was completed in 1877. The first monumental-scale UK structure was the Forth Railway Bridge in Scotland. This was of tubular cantilever construction with a main span of 1710 ft and was completed in 1889. Benjamin Baker's design combined riveted and bolted connections. In 1904 the steel-framed Ritz Hotel in Piccadilly, London was completed, again using a combination of riveted and bolted connections (see Middleton, 1900 for details). The steel was manufactured in Germany and the construction details were in millimetres (perhaps the first significant UK building built using metric dimensions).

The development of structural steel gathered pace at the start of the 20th century. Most structural connections were formed using cleats and rivets.

All ferrous metals corrode in the presence of water and oxygen. In the past structural steel was coated with red lead paint to inhibit corrosion and evidence of this can still be seen today. The main drivers of corrosion (oxygen and water) in combination with steel produce rust. The rate of corrosion is dependent on the time the surface is wet and the presence of atmospheric pollution in the form of sulfates (from sulphur dioxide gas during the combustion of fossil fuels) and chlorides (heaviest in marine environments). Any remedial work must be carefully detailed to avoid entrapment of moisture and dirt. Exposed steel should be treated with an appropriate protective coating applied to a prepared surface free from mill scale and other detrimental material. The protective coating may be in the form of paint or galvanizing. Weathering steels (see also below) should not require protective coatings but should not be used in chlorine-laden environments.

Further research has proceeded to find new and better coatings. One recent example is a product developed by BAC Corrosion Control in association with

Figure 5.20 Wembley Stadium (Courtesy M Wade, Dorman Long Technology).

Mott MacDonald under the title of LATreat. This is a two phase coating developed to tackle the growing problem of accelerated low water corrosion (ALWC). It is claimed that the treatment uses the components of seawater to sterilise affected steel and then deposits a protective calcareous coating.

Of particular interest to practitioners involved in the appraisal and repair of masonry clad steel-framed buildings is the so-called Regent Street Disease in which corrosion of steel is slowly taking place behind the cladding and in some cases has forced off the cladding thus creating a hazardous situation. Reference to the Historic Scotland (2000) publication *Corrosion in masonry clad early 20th century steel framed buildings* will provide useful guidance. In certain circumstances cathodic protection may be used to halt the corrosion of ferrous metals, including structural steel. In dealing with repairs readers are also referred to the section below on welding.

It should also be noted that brittle fracture may be a problem with both new and repaired steel structures. Brittle fracture usually occurs under static rather than fatigue loading and may cause a sudden unheralded failure. Unlike fatigue failures these are not usually preceded by plastic deformation. Gaylord has highlighted the case in 1919 in Boston USA of a 90 ft diameter, 50 ft high riveted steel tank that fractured explosively spilling two million gallons of oil. Fractures usually originate at points of concentrations of stress.

Stainless steel

The quality of stainlessness in steels is provided by the inclusion of at least 12% of chromium. Oxidation produces a dense oxide and adherent film on the surface of the steel which acts as a barrier to further corrosion.

The three main types of stainless steel are:

- martenistic;
- ferritic; and
- austenistic.

For further information the reader is referred to *The Construction Materials Reference Book* (Doran 1992).

Weathering steel

Until the early 1980s this type of steel was usually termed Corten and is a smart metal that forms its own protective barrier under normal atmospheric conditions. It therefore needs no protective coating if used externally under normal atmospheric conditions. With weathering steel the rusting process is initiated in a similar way to other steels but with alloying elements in the material producing a stable water-resistant layer. This rust patina develops under alternate wetting and drying to produce a protective barrier which impedes further access to oxygen and moisture. At the time of writing a major bridge scheme in North Kent was in the process of construction involving 4000 tonnes of weathering steel.

Care should be taken when considering the use of weathering steel in extreme environments which may inhibit the long term durability. In marine environments exposure to high levels of chloride ions in seawater spray, salt fogs or airborne salts are detrimental to this material.

Welding

The process of welding usually involves three factors:

- an electric arc;
- the parent metal; and
- a consumable – a metal alloy.

During the process the parent metal is melted to some depth (known as the penetration). The molten metals (parent and consumable) combine in the weld pool to form a joint. Typical arrangements are fillet welds and butt welds (see Fig. 5.21).

Welding is a highly skilled activity which should only be carried out by properly trained and certified tradesmen. It is important to avoid the introduction of hydrogen into the technique as this may lead to cold cracking. Hydrogen may come from the parent metal, the consumable or the atmosphere.

A recent development is the use of friction welding as used by CORUS in their Bi-steel system for the construction of shear walls and building cores. This technique has been derived from the automotive industry.

For highly stressed work it is important to inspect and test the quality of the weld. Suitable testing regimes are laid down in the National Structural Steelwork Specification (NSSS). (See also section on testing in Chapter 3.) If possible avoid site welding but if it is essential have regard for good access to the weld site.

Ogle (1990) regards 1942 as a watershed in the history of welding. In that year an almost new liberty ship sitting in harbour in calm water had broken its back. Subsequent investigations showed the presence of severe stress concentrations and tensile residual stresses related to welds, which imposed greater than acceptable demands on material toughness. As a result the steel industry produced tougher versions of mild steel.

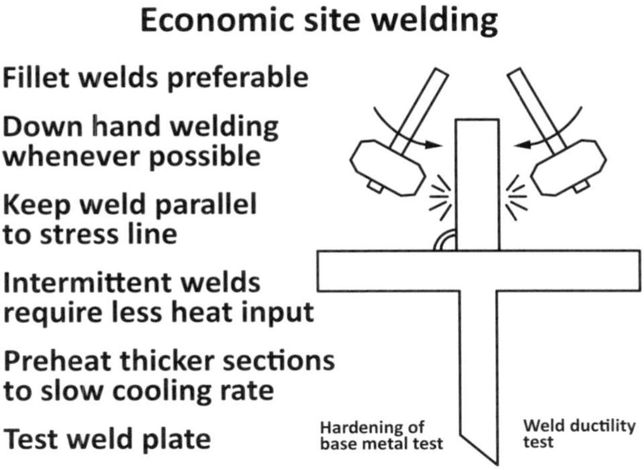

Economic site welding

Fillet welds preferable

Down hand welding whenever possible

Keep weld parallel to stress line

Intermittent welds require less heat input

Preheat thicker sections to slow cooling rate

Test weld plate

Hardening of base metal test

Weld ductility test

Figure 5.21 Welding: good practice (Courtesy CORUS).

In welded connections it is important to see that the details do not allow water to become trapped. It has been known for the freezing of trapped water to burst such connections. Furthermore it is essential to see that rust traps are avoided as rust can exert very high pressures on surrounding steel capable of separating mating sections.

Splash zone phenomena

Air dissolved in sea water increases the rate of corrosion significantly. Dissolved air causes metal surface areas near the sea surface to corrode more severely than surface areas at greater depth. The splash zone extends from some distance below mean low water level (MLW) to about 1.5–2.0 times the distance above MLW. The range depends on local conditions of tide, nominal wave height and (in cold climates) ice abrasion. It is therefore desirable to avoid horizontal bracing to oil rigs (*et al*) in the splash zone. This phenomenon is sometimes referred to as accelerated low water corrosion.

Liquid metal assisted cracking (LMAC)

LMAC is a rare phenomenon that can occur when structural steelwork is galvanised to provide protection from corrosion. Certain solid metals when in contact with other liquid metals can give rise to a reaction which may affect the parent solid material. This reaction is termed liquid metal embrittlement (LME) and may lead to cracking of the steel. For example, when structural steel is stressed and temporarily in contact with liquid zinc in the galvanising process then LME/LMAC may occur.

As with other defect phenomena (e.g. ASR in concrete) a number of factors need to be present to cause the problem. In the case of LMAC three factors have been identified but the critical relative relationship between these factors has yet to be authoritatively established. The factors are:

- stress level
- material susceptibility
- a liquid metal

Examples of LMAC have been identified in Germany, Japan, and America. Research work continues in the UK in an attempt to clarify the issues.

A guide produced by the BCSA identifies areas where the risk of LMAC can be minimised. These include:

- designing and detailing
- type and quality of steel
- quality of fabrication
- galvanizing process

Repair and strengthening

Steel members may be strengthened by adding additional steel sections (plates, angles, channels, rolled steel joists (RSJs) and rolled steel channels (RSCs), etc.) to existing sound members usually by welding. Care must be taken to see that existing and new steel are compatible before welding takes place. Care must also be taken to see that fire resistance arrangements are not compromised. Alternatively, some work has been done by bonding additional steel plates to the original sections. In some applications it will be important to de-stress the original structural member before strengthening. For typical details see Figs. 5.22 and 5.23.

Steel-framed and steel-clad housing systems

- Coventry Corporation steel-framed houses (BR222)
- Riley steel-framed houses (BR221)
- Stuart steel-framed houses (BR219)
- Weir steel-clad (1920s) houses (BR218)
- Cowieson steel-clad houses (BR217)
- Lowton-Cubitt steel-framed houses (BR188)
- Telford steel-clad houses (BR189)
- Cranwell steel-framed houses (BR193)
- Birmingham Corporation steel-framed houses (BR196)
- Hills Presweld steel-framed houses (BR199)
- Arcal steel-framed houses (BR198)
- Homeville industralised steel framed houses (BR199)
- 5 m steel-framed houses (BR200)
- Arrowhead steel-framed houses (BR201)
- British Housing steel-framed houses (BR202)
- Keyhouse Unibuilt steel-framed houses (BR203)
- Open system building steel-framed houses (BR204)
- Steane steel-framed houses (BR205) Nissen-Petren steel-framed houses (BR163)

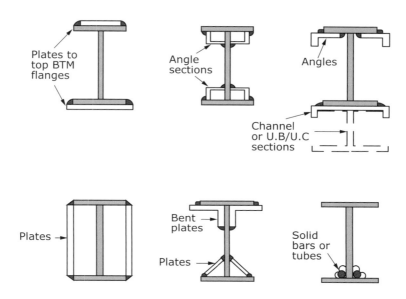

Figure 5.22 Strengthening techniques for steel beams (1) (Courtesy CORUS).

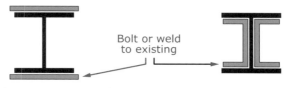

Site welding to existing steels

EXISTING STEEL	COMMENTS
Cast iron	Cannot be welded (except non structural fixings - very high nickel electrodes)
Wrought iron	Low hydrogen electrodes. (Dry to manufacturers recommendations (approx. 350°C). Sections > 25 mm preheating required.
Mild steel Low carbon < 20 High sulphur	Low hydrogen electrodes. Sections > 25 mm preheating required.
Mild steel High carbon (1910–1930 steels)	As above but preheating required for all structural welds.

Figure 5.23 Strengthening techniques for steel beams (2) (Courtesy CORUS).

- Cruden rural steel-framed houses (BR139)
- Falkiner-Nuttall steel-framed houses (BR144)
- Crane steel-framed bungalows (BR145)
- Trusteel MK11 steel-framed houses (BR146)
- Trusteel 3M steel-framed houses (BR147)
- Atholl steel (BR148)
- Hawthorn Leslie steel-framed houses (BR152)
- Roften steel-framed houses (BR119)
- Dennis-Wild steel-framed houses (BR120)
- Cussins steel-framed houses (BR132)
- Livett-Cartwright steel-framed houses
- Dorlonco steel-framed houses (BR110 and BR149)
- Thorncliff cast-iron panel houses (BR111)
- BISF steel-framed house (BR77)
- Howard steel-framed house (BR78)

5.2.4 Timber

See also Section 5.3.6.6 'Fire in timber structures'.

5.2.4.1 General

Timber has been used in construction for many hundreds of years. Naturally occurring and constantly replaceable, it has the essential merit of being easily workable. One of its endearing qualities is that it is normally a naturally renewable material. One of the earliest examples of timber housing is the medieval Cruck house of which a few still survive. In this system the roof is carried on pairs of timbers from ground level to ridge.

Timber is generally classified into two categories – softwood and hardwood. These are botanical distinctions unrelated to physical properties (see Appendix). Typical softwoods might be pine and spruce; hardwoods, ash and oak.

The standards against which forestry management techniques can be measured are laid down in the UK by the Forest Stewardship Council (FSC).

Modern timber framed housing makes maximum use of off-site fabrication. Pre-fabricated panels are brought to site and speedily erected to form a watertight enclosure in less than a working day. These panels are provided with plasterboard, vapour barriers and initial fixings for service runs, thus reducing the need for on-site work. Interest in timber framed housing was generated in the 1970s by increased thermal insulation requirements. In 1974 over 23,000 houses were built in this way. By 1980 the number had risen to 46,000 and although there have been setbacks due to doubts about quality, timber framed housing has continued to occupy a considerable sector of the market.

Structural timber is typically Canadian spruce or hem-fir, European redwood or European whitewood, pressure treated with copper/chrome/arsenic (CCA) preservative. Fire resistance is provided with the appropriate number of layers of plasterboard to suit location. Roof structures are typically trussed rafters jointed using galvanised gang nail plates. External elevations are usually single brick or block skins with cavity separation from the timber frame. Problems that have arisen have been mainly caused by poor on-site storage that has permitted prefabricated panels to become wet and affected by mud from poorly organised storage areas.

5.2.4.2 Natural defects

- Fissures: a term covering all splits, checks and shakes that result from a longitudinal separation of the wood fibres due to shrinkage caused by the timber drying and may appear on any face, edge or end of the timber.
- Wanes: the original rounded surface of a tree remaining, with or without bark, on any face or edge of square sawn timber.
- Knots: these usually occur on a tree at the junction of a branch with its parent member. Knots may work loose or otherwise reduce the strength on a timber component.

5.2.4.3 Building fungi and wood rot

Wood-rotting fungi are frequently responsible for the deterioration of timber. In particular, elm is badly affected by *Dutch elm disease* in which elm bark beetles spread disease from tree to tree. Generally, this deterioration is unlikely in timber with a moisture content of less than 20%. Wood-rotting fungi get their food by breaking down the wood cell walls, thus causing loss of strength. Not all fungi found in buildings cause wood rot. High moisture content often has its origins in damaged or non-existent damp-proof courses and other waterproofing membranes, broken or inadequate rainwater drainage, poor ventilation, damaged or defective flashings.

Particular areas of vulnerability include: joists of suspended floors adjacent to brickwork supports, wall plates supported on sleeper walls that are devoid of damp proof courses, roof timbers adjacent to damaged roof coverings or inadequate flashings and joists of suspended timber floors that are inadequately ventilated.

In refurbishment work particular attention should be paid to the selection of appropriate timber and the preservation of new timber by chemical impregnation and other methods. Such defects may also be caused by water leakage from faulty services. Ideally, timber should be selected from renewable sources.

5.2.4.4 Dry rot

Dry rot is caused by the fungus *Serpula lacrymans* and is more prevalent in softwoods rather than hardwoods. This condition may need to be dealt with by cutting back the affected section and replacing a part or all of the structure. As a rule of thumb, it is prudent to cut back to at least 1 m beyond the affected area. It is also prudent to check that the location in which the problem occurs is adequately ventilated and the source of the damp eliminated.

5.2.4.5 Wet rot

Wet rot is caused by wood-rotting fungi other than *Serpula lacrymans* (e.g. *Coniophora puteana*) which characteristically attack comparatively wet timber in structures. This condition has traditionally been dealt with by drying out the affected area. If the attack is not too severe and if the residual member cross-section is structurally sufficient then little more needs to be done. If the attack is severe it may be necessary to augment or replace the structure. At a moisture content of 25% or above fungal attack is likely. It is advisable to keep structural timber to a level below 20%.

5.2.4.6 Insect infestation

A number of insects [mainly beetles] may attack timber in search of food. In extreme cases this may affect the capacity of the timber and require remedial treatment. Bravery *et al.* (2003) have classified damage categories *A*, *B*, and *C* in descending degree of seriousness: *A* usually needing insecticidal treatment; *B* treatment only necessary to control associated wood rot; *C* no treament required.

In temperate climates, where timber is maintained in a dry condition, insect attack will not usually occur. There is one exception, however, which occurs in the presence of the House Longhorn Beetle. The presence of this insect is largely confined to some parts of southern England, specified in UK Building Regulations. In these areas the use of preservative treatments in roof timbers is mandatory.

5.2.4.7 Defective jointing (including breakdown of glued joints)

Joints in timber structures may be defective due to poor detailing and construction or a breakdown in glue or other adhesive. In older properties carpenters' joints may be present in structural assemblies, such as timber roof trusses. If, due to shrinkage, these joints have opened up it may be possible to strengthen them by the use of glue. Alternatively the joints may be strengthened by the application of metal or plywood gusset plates screwed and/or glued to the original structure.

The authors are aware of the collapse, some years ago, of a series of timber roof trusses in which the joints were formed using plywood gusset plates. These joints had failed due to the use of sub-standard glue. Where glue is used it should be supplied according to the appropriate BS and properly applied. In the case of trussed rafters, where analytical structural analysis is uncertain, load-testing should be used to determine capacity.

In cases where joints in timber structures rely on carpentry joints these should be made to a sufficiently tight fit and can only be successful if executed by appropriately skilled carpenters.

5.2.4.8 Metal corrosion

It should be noted that embedded metals such as lead, iron and steel may corrode in damp timber such as oak. Some corrosion of metal in wood may be caused by the type of preservative or flame retardant used.

5.2.4.9 Repair and conservation

Christopher Mettem of the Timber Research and Development Association (TRADA), in a recent paper, reports on innovative materials and techniques in the conservation and repair of timber structures.

The International Council on Monuments and Sites (ICOMOS) has recommended that practitioners should:

- Plan and conduct interventions that maintain historical authenticity and cultural integrity.
- Put into effect repairs that respect historical, aesthetic and scientific values.
- Recognise the importance of timber from all periods as part of the cultural heritage of the world.
- Consider the great diversity of historic timber structures and of the species and qualities of wood used to build them.
- Recognise the vulnerability of these structures and their increasing scarcity due to the nature of timber and the loss of skills and knowledge in design and craftsmanship.

The paper deals principally with the use of epoxy resin as a material to repair or enhance the structural capacity of timber members. It highlights the battle to convince practitioners that these methods were as acceptable as previously used carpentry-based solutions. Individual projects studied by TRADA included:

- Blackfriars, Gloucester South Range (repair and conservation of a scissors trussed oak roof, dating from 1239, a Scheduled Ancient Monument).
- St James the Great Church, Essex.
- Nashs's 1820–1821 Theatre Royal, Haymarket, London (an iron-connected pine roof truss structure also involving stage machinery.)

Mettem reports on the international Glued-in Rods for timber (GIROD) project set up to provide information required for standards and design rules allowing an increased, more advanced and more reliable use of bonded-in rods in timber structures. The following summary indicates progress with this research programme:

- Design rules are available for Eurocode 5. A calculation model is based on Volkersen theory supported by fracture mechanics, giving good prediction of anchorage strengths for the appropriate failure models.
- Guidance now exists on spacing distances between rods and end and edge distances.
- The effect of moisture is known and is shown to depend upon the adhesive type. Adjustment factors have been determined.
- Rules are given for duration of load effects: UK faith in epoxy types was justified by the tests.
- Fatigue limits might apply in certain structures (mainly bridges). Suitable design rules have been derived.

- Test methods have been developed for adhesives. One is capable of ranking adhesives for durability, whilst another determines the creep-rupture behaviour of small bonded-in rod specimens, supporting the design expressions mentioned above.
- A proof loading test method for production control has been developed, and shown to be capable of detecting common production defects.

Work also continues through the Low Intrusion Conservation Systems (LICONS) international research project to develop less intrusive repair methods using, for example, wide gap-filling adhesives in combination with fibre-reinforced polymers (FRPs). Benefits claimed include:

- Low disturbance: ceilings and beam soffits retained; floors and roofs remaining intact.
- Building occupancy: reduction in exclusion with smaller closed areas and shorter possession times.
- More conservationally acceptable: retention of existing timbers, sizes, species, qualities.
- Authenticity: replacement, where essential, on a like for like basis as far as possible.
- Durability: inbuilt protective design; inclusion of features such as end-grain sealing and additional ventilation.

5.2.4.10 Repair and strengthening

Timber beams may be strengthened using steel angles or channels bolted to the sides of the existing beams. As an alternative, steel flitch plates may be inserted into slots cut in the existing timber and either bolted through or used in conjunction with epoxy resin bonding agents. For details see Fig. 5.24.

Alternative methods involve creating flitched beams using additional timbers bolted through the original timber. For additional information see Carmichael (1984).

5.2.5 Glass

5.2.5.1 General

Glass is a very old material dating back to 10,000 BC in Egypt. The common glass used in windows is usually soda glass which is made by heating a mixture of lime (calcium oxide), soda (sodium carbonate) and sand (silicon dioxide). Glass used for structural purposes (e.g. for fins to stabilise large glass panels) will command a higher specification. Glass to resist fire may well contain at least 5% of boron oxide B_2O_3.

Types of glass encountered in building construction include:

- annealed
- laminated
- toughened

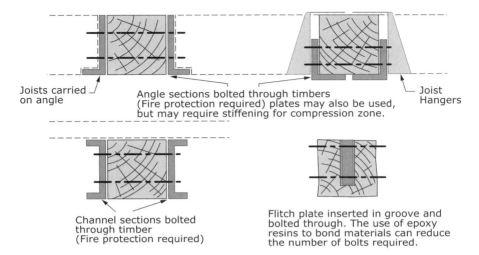

Joists carried on angle

Angle sections bolted through timbers (Fire protection required) plates may also be used, but may require stiffening for compression zone.

Joist Hangers

Channel sections bolted through timber (Fire protection required)

Flitch plate inserted in groove and bolted through. The use of epoxy resins to bond materials can reduce the number of bolts required.

Figure 5.24 Strengthening structural timber (Courtesy CORUS).

- wired
- fire resistant (see also Section 5.3.6.7 'The effect of fire on glass'.)

Glass is also fashioned into glass blocks (e.g. for pavement lights) and features as structural glazing. Proportions will vary but a guide to the specification of typical soda glasses can be gained from the following:

Silicon dioxide 69–74%
Calcium oxide 5–12%
Sodium oxide 12–16%
Magnesium oxide 0–6%
Aluminium oxide 0–3%

A full description of each type is beyond the scope of this book but reference may be made to *Glass in Buildings* (Button and Pye 1993) for more detailed information.

Modern glazed facades often consist of glass panels stiffened by glass fins glued with structural silicones to the facing glass. It is important when considering refurbishment to recognise the structural interaction of facing and fin glass. Damage to the fins may impair the structural stability of the system. In alternative forms of construction glass is glued to aluminium framing.

Glass has excellent resistance to salt water, strong acids, organic solvents and aerated water but poor resistance to strong alkalis. In terms of embodied energy it is the middle of the range at 32–57 GJ/m^3.

In spite of its brittle and unpredictable nature, glass can be used to resist blast, hydrostatic pressure and impact. Whilst selection of material specification is important due allowance for these situations must be borne in mind when designing supporting frames and fixings.

Where coated glass is used for reflective or other purposes it should be noted that the coating may have a shorter life than the glass. This should be taken into consideration in assessing replacement periods.

With advancing technology glass has become a very sophisticated material as a result of which it can be produced with characteristics which include:

- security (e.g. bullet proof, blast resistant, anti-bandit proof)
- fire resistance (see also Section 5.3.6.7 'The effect of fire on glass'.)
- solar control
- thermal performance enhancement

5.2.5.2 Fixing problems

As with all construction materials, glass is subject to thermal and structural movement. Allowance for these factors must be made when designing support frames and fixings. In glass assemblies reliant on bolted fixings the glass must be insulated from bolt stress by the used of suitable gaskets and washers. Edge distances to fixing bolts and their attendant drilled holes should strictly adhere to manufacturers' recommendations to avoid premature failure of the glass.

5.2.5.3 Nickel sulphide inclusions

A nickel sulphide inclusion is formed when nickel (usually from the sand) and sulphur (from the fuel) combine. Inclusions are often small (perhaps 100 μm) but have been known to be up to 4 mm. These expand under the action of light (and/or differential temperatures) and exert local stress on the glass. If within or close to the tensile zone of toughened glass then critical stress levels may be reached and cause shattering as energy is released. In the 1980s (e.g. in Jersey) several failures occurred in toughened glass: they are less likely in non-toughened laminated glass where built-in stress levels are normally lower.

Obviously the only solution to an individual problem is to replace the glass but only after a close examination of the surrounding environment to check if an alternative glass specification (such as heat-soaking or laminated) can be accommodated.

5.2.5.4 Thermal shock

Sudden changes of environment (e.g. rapid rise in temperature due to a burst of heating from modern air conditioning units or summer hailstorms) may cause glass to fracture.

5.2.5.5 Weld spatter

Weld spatter (sometimes called weld splatter) occurs when small particles of hot metal land on the surface of the glass and penetrate the surface before cooling. This can happen when adjacent welding or metal grinding is taking place close to the glass. The embedded particles of metal corrode with time; the corrosion products exceed the original size of the metal fragments thus causing high internal stresses in the glass and subsequent fracture. In inspection of existing glass it is important to distinguish weld spatter from accumulated dirt.

5.2.5.6 Fatigue loading

The creep capability of glass, unlike concrete, is low. The material therefore has a poor ability to absorb long term movements.

5.2.5.7 Edge shelling

This can occur when due to careless handling, transportation or poor installation practice the finished edges of glass panels are damaged. The damage caused by poor installation may take a long while to occur. The absence of support material can result in progressive movement of glass within a frame and eventually lead to contact between glass and metal with resultant flaking of glass from an edge long after the original installation.

5.2.5.8 Scratch damage

Scratches deeper than 400 μm (detectable by a fingernail test) may cause failure. Such damage may have been caused by window cleaners abseiling tackle and care must be taken to provide protective cover to such equipment.

5.2.5.9 Other problems

- Deterioration of film-protected glass such as that used for anti-blast protection. Recommended film thickness should not be less than 175 μm and extend to within 1 mm of the edge of the pane.
- Inadequate allowance for movement. Tolerances should be limited to ±2 mm. Gasket arrangements should provide a good seal but also allow for adequate movement.
- Deterioration of seals on double glazing units

See also Section 5.3.4.7, 'Windows and doors'.

5.2.6 Polymers (plastics)

It should be noted that it is not normally feasible to strengthen or repair plastics. See also Section 5.3.6.8 'Fire in Plastics'.

5.2.6.1 General

There is a misconception that all polymers are man-made. However, this is untrue. For example rubber, which has its origins in the sap from a tree, is a natural polymer. Synthetic polymers may be categorised into two types; thermosets (e.g. Bakelite) and thermoplastics (e.g. polystyrene). Polymers include:

- Natural: rubber tapped as latex from trees and then usually vulcanised; celluloses occurring in plant life including timber form the basis of lacquers for wood finishers; also mouldable into rigid materials sometimes

Table 5.5 Typical plastics failures.

Defect	Likely cause	Remedial action
Burning of surfaces	Fire – resulting in off-gassing of toxic fumes. Acid attack. Burning droplets.	Replace defective sections.
Discolouration	Soiling/staining. Smoke damage. Pigmentation loss. Ultra-violet radiation.	Clean with an approved pvc chemical cleaning agent. Thereafter consider coating it with a compatible polymeric paint.
Disintegration	Plasticiser migration from exposure to bitumen.	Keep plastic and bitumen apart using inert separating layer.
Embrittlement	Chemical attack. Ultra-violet radiation.	Replace affected sections.
Erosion, scratching and scoring of surfaces	Persistent or abnormal wear and tear.	Replace affected sections.
Indentation or breakage	Impact or other mechanical damage.	Replace affected sections.
Softening	Exposure to excessive heat – usually triggered by infra-red radiation.	Replace affected sections.

used for shatter-proof glazing and hand rail coverings also used as pre-formed water bars although this has now largely been overtaken by the use of PVC.

- Synthetic – thermosets: often known as Bakelite, can be easily extruded to make electrical components due to its good insulating properties.
- Synthetic – thermoplastics: e.g. polystyrene used for void-forming in concrete; as uPVC in window and door frames; as acrylic sheeting for sanitary ware; gas and water supply pipes and as damp proof membranes.

Despite their general durability, plastics, like most building materials, are also not immune to deterioration. They are vulnerable to attack by a number of mechanisms, as listed in Table 5.5.

5.2.7 Other materials of interest

Other authorities have identified more than 50 materials that are used in construction. It is not within the scope of this book to deal with all of these but useful guidance is supplied in other publications given in the Bibliography and further references.

5.2.7.1 Wattle and daub

In medieval timber-framed buildings wattle and daub was the traditional method used to fill the space between the large timbers that make up the structure of the building. The use of this material is thought to date back to Neolithic times and there is evidence of its use in Ireland as early as 6000 BC. Its use has been confirmed in Central Europe, Western Asia and America. With sustainable construction now high on the agenda its use may make a modest revival.

The wattle is made from small timber sections fixed between the structural timbers and the daub is then applied to give a finished surface on both sides of the wall. In the best quality work vertical oak staves were fixed to the timber structure then thin clefts of oak (i.e. split lengths of oak about 25 mm × 2–3 mm thick) were woven between the staves. These panels are then daubed with a mud and straw mix on both sides giving a reasonable surface inside and outside the building. This surface is adequate to receive a lime wash finish and will keep the wind and rain out of the building.

The daub is made from earth (subsoil is the best), mixed with straw or grass. Sometimes lime or cow dung is added to give greater flexibility and workability.

Availability of materials affected the way the work was carried out. Locally available timber would determine what was used. Cheaper work would make use of twigs and branches rather than prepared timber. The end result is an infill panel made from locally sourced, renewable materials, with almost zero energy use, flexible both in the form in which it is applied and its ability to flex as the timber frame dries out. Maintenance would be a fairly frequent requirement as cracks occurred and water caused the mud to break up.

Examples of the way wattle and daub is used in the reconstruction of timber framed buildings can be seen at the Weald and Downland Open Air Museum in West Sussex. Further information is available on the museum's website (see Appendix, Section A2.3).

5.2.7.2 Naturally sourced materials

Increased demand for sustainable and renewable components is leading to renewed interest in naturally resourced materials. Whilst refurbishment currently offers little scope for consideration of these materials their importance is likely to increase. These materials include straw bales, rammed earth and bamboo.

Straw bales

It is claimed that farm buildings constructed using load-bearing compressed straw bales has been in use in Nebraska USA for over 100 years – some of which are still in existence. Recording studio buildings have also been constructed in the USA on the basis that the bales provided good sound insulation. The use of the technique in the UK has grown steadily since 1995 and examples exist of housing and studio type buildings.

Straw bales can be made of seed- and weed-free wheat, barley, oats and perhaps of rye or hops. The preferred sizes are 360 mm × 460 mm × 920 mm or 41 mm ×

58 mm × 114 mm with a maximum moisture content of 15%. Rectilinear bales are formed using ties of sisal, baling wire or polypropylene. Units placed on traditional foundations are assembled as large, coursed blocks, pinned vertically with bamboo, coppiced hazel or softwood spikes. Compressed straw in fire chars in a similar manner to large cross-section timber and has been fire tested in Australia and Canada. UK designs have satisfactorily complied with building regulations.

Advantages claimed for these systems include good sound and thermal insulation; easy handling (bales are roughly only one third of the comparative weight of brick or concrete blockwork); the need for only semi-skilled or unskilled construction labour and easy recycling (composting) at the end of useful life. Walling can be readily shaped to accommodate windows and doors and can be rendered with lime mortar then decorated in the traditional manner.

In an alternative form of construction straw bales are used as cladding or in-fill to steel or timber structural frames.

Bamboo

Bamboo is a species of grass, not wood. It is fast growing and reaches maturity in about 3–5 years at a height of 20–25 m and, if left untouched, would have a 20-year life. There are in excess of 1100 species and of these 450 occur in the Americas. In South America it is found in Colombia, Ecuador, Brazil and Venezuela and is still being introduced. Of particular interest to the structural engineer is the *guada* strain. Typically a bamboo has no cambium, bark or radial fibres and grows to full height at an approximately constant diameter.

The mechanical properties of bamboo highlight its superiority to a C16 softwood in compression, axial tension and bending strength. However it performs badly in shear, and in tension perpendicular to the grain. Unprotected, its durability and fire resistance is poor although its strength to weight ratio exceeds that of steel.

Historically there is evidence of the material being used structurally for short span bridges and housing in the 19th century with examples of the latter still in use after 100 years. There are excellent examples of low-budget housing where the bamboo framework has been sheathed and protected by a coating of cement mortar. Performance in earthquakes is extremely good, and compares favourably with the performance of comparable reinforced concrete structures.

A residual difficulty was that of achieving satisfactory structural connections. Solutions tried included the use of lashing, nails, screws, bolts, dowels and pegs. More recently research projects had been commissioned to examine the benefit of locally filling the bamboo with cement grout to achieve a more efficient solution. Typical in all types of connections were splitting and bearing failures.

Entrepreneur Simon Velez had given South Americans inspirational leadership in developing bamboo as a major construction material. In housing, savings in the order of 40% were achievable in comparison to the use of masonry construction. This process was being encouraged by the development (in South America) of design codes and it was not unreasonable to envisage buildings of 4–5 storeys for the future.

In the UK bamboo has been used for temporary structures and also as part of decorative cladding systems. Repair usually takes the form of replacement.

5.3 Other matters

5.3.1 Adverse environmental conditions

5.3.1.1 Earthquakes

An earthquake is a vibration, sometimes violent, of the earth's surface (see Fig. 5.25). It is usually caused by tectonic plate movement but may be caused by volcanic activity, explosion, collapsed mine workings or water pressure in reservoirs. Where a considerable disruption of sub-sea tectonic plates has occurred it is likely that a *tsunami* will occur adding severe flooding to areas already devastated by the earthquake shock. There are two traditional measures to describe earthquakes:

- The Richter Scale (see Table 5.6) formulated in 1935 by Charles Richter of the California Institute of Technology (CIT). This is a logarithmic scale (to the base 10) that records the magnitude of an event in the range 1–10. For example an earthquake which registers 3.0 on the scale would be described as minor; it may be felt but would rarely cause damage. One recording 7.0 might cause serious damage over a wide area. The largest recorded event to date was in Chile in 1960 which registered 9.5 on the Richter scale.
- The Modified Mercali Intensity (MMI) scale (see Table 5.6) was developed in 1931 by Harry Wood and Frank Neumann in America. This records the effect of an earthquake on a scale in Roman numerals of I to XII. For example MMI II would be felt by a few people at rest; delicately suspended objects may sway: An event measuring MMI IX would cause considerable damage even in specially designed buildings with structures out of plumb and possibly dislodged from their foundations.

Figure 5.25 Eastern Europe: effect of severe earthquake on office building.

Intensity (MMI)		Effect	Approx. equivalent magnitude (Richter)
I	Instrumental	Not felt; detectable only by a seismograph.	1–3
II	Feeble	Felt only by sensitive persons or those at rest.	
III	Slight	Felt indoors, possibly only by those at rest; hanging objects swing. Vibration comparable to the passing of a heavy vehicle.	
IV	Moderate	Felt; hanging objects swing; standing motor cars rock; windows, doors etc. rattle. In the upper range of IV wooden walls and frames creak.	4–5
V	Rather strong	Felt outdoors; sleepers wakened; suspended objects swing, loose objects fall.	
VI	Strong	Felt by all; causes alarm; trees sway; pictures etc. come off walls; windows may break; weak plaster and masonry type D may crack.	
VII	Very strong	Difficult for persons to stand; general alarm; walls crack; plaster fails. Damage to masonry D, including cracks. Weak chimneys broken at roof line, fall of plaster, loose bricks, stones, tiles, cornices (also unbraced parapets and architectural ornaments). Concrete irrigation ditches damaged. Some cracks in masonry C.	
VIII	Destructive	Steering of cars affected; branches broken from trees; damage to masonry C, partial collapse. Some damage to masonry B. Fall of stucco and some masonry walls. Fall of chimneys, factory stacks, monuments, towers, elevated tanks. Frame houses moved on foundations if not bolted down; loose panel walls thrown out. Decayed piling broken off.	6–7
IX	Ruinous	General panic; some houses collapse; ground cracks; pipes break. Masonry D destroyed; masonry C heavily damaged, sometimes with complete collapse; masonry B seriously damaged; frame structures, if not bolted down moved off foundations.	
X	Disastrous	Many buildings destroyed; ground cracks badly; railway lines twisted; landslides on steep ground. Most masonry and frame structures destroyed. Some well-built wooden structures and masonry destroyed.	
XI	Almost catastrophic	Few buildings remain standing; bridges destroyed; landslides and flooding (due to destruction of reservoirs, dams, embankments, pipes). Rails twisted greatly. Underground pipes out of service.	8–10
XII	Catastrophic	Total destruction. Objects thrown into the air. Large rock masses displaced. Lines of sight and level distorted.	

Masonry A: Good workmanship, mortar and design; reinforced; especially laterally, and bound together by using steel, concrete, etc.; designed to resist lateral force.
Masonry B: Good workmanship and mortar; reinforced, but not designed in detail to resist lateral forces.
Masonry C: Ordinary workmanship and mortar; no extreme weaknesses such as failing to tie at corners, but neither reinforced nor designed against horizontal forces.
Masonry D: Weak materials, such as adobe; poor mortar; low standards of workmanship; weak horizontally.

Facing: Table 5.6 **Earthquake scales:** the Modified Mercali Intensity (MMI) scale and the approximate equivalent magnitudes of the Richter scale (based on P.G. Fookes *et al.*, 2007).

In spite of popular opinion to the contrary earthquakes do occur in the UK. For example on 23 September 2002 an event measuring 4.8 on the Richter Scale happened in Dudley, West Midlands with some damage to domestic properties. It is not without significance that this was an area of extensive coal mining activity. More recently in April 2007 an event which measured 4.3 on the Richter scale occurred in Folkstone, Kent. Fortunately there were no deaths but some residents were injured when the earthquakes cracked walls and demolished chimneys.

When dealing with the repair of structures subjected to earthquake damage practitioners would be well advised to seek expert advice.

5.3.1.2 Strong winds

Until the 1970s most UK designs for wind loading were carried out using BSCP3 Chapter V (1952). In the 1970s, in particular in BSCP3: Chapter V: Part2: 1972, a more sophisticated approach was taken. For the first time maps showing basic wind speeds for the UK began to appear and more attention was given to the local topography of sites. This trend has continued with new emphasis on, for example, such items as dominant openings in buildings such as occurs in aircraft hangars where a series of doors might occupy a complete elevation (see also Section 5.1.1: 'The Great Storm in South East England, 1987'). A popular way of describing winds is in the use of the Beaufort Scale (see Table 5.7).

Table 5.7 Beaufort wind scale.

Beaufort Wind Force	Wind Speed m/s	Description	Sea description
0	0–0.2	Calm	Like a mirror
1	0.3–1.5	Light air	Rippled
2	1.6–3.3	Light breeze	Small wavelets
3	3.4–5.4	Gentle breeze	Large wavelets
4	5.5–7.9	Moderate breeze	Small waves
5	8.0–10.7	Fresh breeze	Moderate waves
6	10.8–13.8	Strong breeze	Large waves
7	13.9–17.1	Moderate gale	Sea heaped up with white foam breakers
8	17.2–20.7	Fresh gale	Moderately high waves with spindrift
9	20.8–24.4	Strong gale	High waves with streaks of foam
10	24.5–28.4	Whole gale	Very high waves with overhanging crests
11	28.5–32.7	Storm	Exceptionally high waves with extensive foam patches
12	>32.7	Hurricane	Sea completely covered with foam, driving spray

5.3.1.3 Snow loads

In the 1980s there were some incidents of damage to the roofs of industrial buildings in which a roof slope was adjacent to a vertical or near vertical surface. This allowed excessive depths of snow to build up and overload the roof. Similar conditions might also apply in roof valleys or local projections and obstructions. This phenomenon was recognised in an amendment to the relevant BRE literature and BSI loading codes.

5.3.1.4 Climate change

For a number of years engineers and others have been examining the potential effects on buildings and structures of projected changes in the climate. Although there is still a great deal of controversy over this topic it is now the considered view of the Stern Committee that human behaviour contributes to climate warming and its concomitant effects. It is accepted by SCOSS (2000–2001) that the most likely hazards to structural safety arising from climate change are:

- extreme winds
- extreme precipitation, especially snow
- extreme precipitation leading to flooding and scour (see also Bridges in Section 5.3.24)
- periods of drought and high temperatures leading to ground movements
- extreme depositions of ice on structures
- extreme diurnal temperature changes
- more severe wave action at sea

To this it might be reasonable to add rises in sea levels.

Practitioners would be well advised to consider these issues especially in respect of buildings or structures requiring a long design or residual design life.

It is now generally accepted that greenhouse gas emissions (such as carbon dioxide) are instrumental in affecting climate change. The approach to this of the UK government has been to attempt to cap total carbon emissions (1.1 MtC/year by 2020) and allow within those limits for organisations to trade with others to achieve those limits. Thus an organisation in danger of exceeding the target may pay to have beneficial trees planted in another country. Alternatively such an organisation can financially assist another that can achieve with ease the target emissions. Practitioners involved in refurbishment, repair or reconstruction should make allowances for this situation.

5.3.1.5 Flooding

Possibly due to climate change, severe flooding has become more prevalent in the UK. There has also been a dangerous tendency to disregard flood plains in some recent planning decisions. Research is now underway to improve flood protection particularly in coastal areas. Where the cost of defences is excessive it has been decided, for example in parts of East Anglia, to allow land to flood.

Meteorologists have suggested that the main contributory factor leading to the 2007 floods was the displacement of the Gulf Stream to a level 1500 miles south of its normal position.

A growing population may demand that more housing be built in vulnerable areas. Some new thinking is needed and recourse to Dutch research may be enlightening.

5.3.2 Condensation

Most air contains water vapour and when the air becomes saturated with the vapour some will condense out as water. Warmer air is able to hold more water vapour than air at a lower temperature. In buildings, condensation typically occurs when warm moist air meets a cold surface (e.g. run a bath and the mirror steams up).

In some situations the moisture from the condensing vapour will be visible on the surface of the building fabric. The condensation could also occur within the fabric of the building and only be evident by touch or by testing.

If the fabric of a building is damp it will be colder than adjacent dry material and therefore more prone to condensation.

The fabric of a building is the key to condensation control. Thermal insulation, heating, ventilation and possibly cooling need to be considered as part of the total design.

But to avoid condensation a balance is needed between the way a building is used and the way the fabric works. The way it works will be affected by the construction, the condition of the fabric, thermal performance and defects, mechanical and electrical systems including heating and ventilation.

5.3.2.1 Types of building

The use of a building will determine the amount of water vapour produced. The way the building works will determine the effects of that water vapour on the fabric. Sources of water vapour include:

- people breathing out and perspiring
- activities within domestic buildings (cooking, bathing, clothes drying, some forms of heating)
- activities within non-domestic buildings (vapour from swimming pools, activities in offices, activities in indoor sports areas, vapour from industrial processes)

The way the building works is affected by:

- the thermal insulation of the fabric
- the internal and external temperature
- the amount and type of heating, ventilation and extraction
- the routes for moisture to move around the building

For buildings such as those containing swimming pools and industrial processes the requirements and design will be specialised and detailed. These types of building will have a known level of vapour production and require controls and extract systems designed to deal with those levels. In domestic buildings the process is more difficult to predict. The production of vapour is determined by the occupants and the way they live. The occupants will also affect the amount of heating, ventilation and extraction as well as the routes for moisture to move around the building.

5.3.2.2 Principals of condensation control

- Avoid the production of moisture (avoid the use of paraffin and unvented bottle gas heating, microwave food instead of boiling).
- Prevent the moisture entering the air (put lids on pans when boiling, vent tumble dryers to the outside air).
- Prevent the moisture spreading around the building (close the kitchen door).
- Extract at source (use the extract fan when cooking).
- Prevent moisture entering vulnerable rooms by closing doors (especially to unheated bedrooms).
- Allow moisture to escape (open trickle vents and windows).
- If possible leave heating on at a low level overnight (low overnight temperatures will lead to more moisture condensing from the air).
- Kill and remove mould growth (use fungicidal wash or a weak bleach solution if the surface is suitable).
- Allow air to circulate (leave at least a 50 mm gap between vulnerable walls and furniture).

5.3.2.3 Changes to the fabric

Changes made to the fabric of a building during refurbishment will alter the response to condensation. An improvement in thermal performance for one element of the fabric might create a condensation problem elsewhere that did not exist before.

An example of this would be the replacement of old draughty windows with well sealed double glazed windows. As a result the air changes in the building might be reduced from 3 or 4 changes an hour down to below 1 change per hour. This in itself could be enough to create severe visible condensation and mould growth.

Condensation will be attracted to the coldest surfaces first. The glass of a single glazed window is likely to be the coldest surface in a room. If condensation occurred on the glass window there would be a reduction in the amount of moisture in the air. With double glazed windows the glass is unlikely to be the coldest point in the room and condensation on the glass is unlikely. The result of this is an additional amount of vapour in the air to condense out in other places. Unless improvements have been made to the thermal performance of the walls the new cold spot could be above the window.

If the thermal performance of walls and roofs is to be improved it is essential to know the details of the existing construction. The basic options for improvement are:

- external insulation;
- internal insulation; and
- improvements to the wall itself.

Changing the insulation of a wall will have an effect on the way it responds to water vapour. Some older forms of construction can be permeable to water vapour (e.g. emulsion paint, lime plaster, solid brickwork). As the vapour passes from the warm moist air inside the building to colder air outside, vapour could be released as moisture within the fabric. Increase the level of insulation and the risk of this interstitial condensation increases. The likely result of this needs to be calculated and the risk controlled.

Preventing the vapour entering the wall is the usual action. A vapour check on or near the inner surface of the wall will achieve this. Other materials closer to the outside of the wall construction could provide some resistance to vapour movement. These must have lower resistance to vapour and allow sufficient vapour through to prevent trapping moist air within the wall. Detailed changes to the fabric can also have an effect. If the wall finish and construction allows water vapour to pass through there is less chance of surface condensation occurring. If an impervious paint or paper is then applied to the wall surface the chance of condensation occurring on the surface is increased. If a wall suffers from surface condensation, providing an impervious surface finish will not solve the problem.

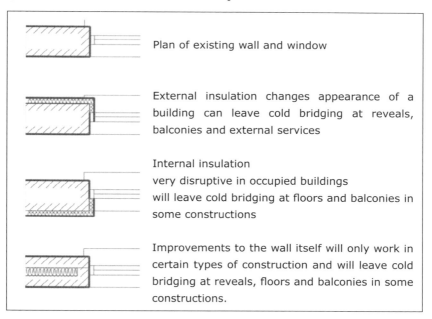

Figure 5.26 Thermal performance improvements.

5.3.2.4 Fabric and saturation

Some building materials can survive when damp but in this state they will transfer moisture to dryer materials unless damp proof membranes provide separation. Most building materials will decay if damp or will cause contiguous materials to decay.

The first visible signs of this process could be deterioration of finishes. The hidden fabric could also be suffering considerable damage: gypsum plaster loses strength; timber becomes prone to dry rot, wet rot and insect attack, warping and twisting; ferrous metals rust and expand; adhesives become less effective leading to finishes lifting.

Protecting a structure from saturation during construction and repair is very important. If, say, the roof of a building is removed and the walls become saturated due to rain penetration the drying out time is likely to be about 25 mm a month for the wall thickness. If the building is occupied during this drying out process the risk of condensation occurring will be increased.

5.3.2.5 Testing and prediction

The affects of changes to thermal performance of the fabric can be calculated. Calculations can show areas that are at risk from condensation and where new problems could occur. Despite the affects of lifestyle on a dwelling, this process can provide an accurate guide to what will happen – you can look at the figures predicting the mould growth and then look at mould growth on the wall.

$$\text{Temperature difference ratio (TDR)} = \frac{\text{internal air temperature - cold bridge temperature}}{\text{internal air temperature - external air temperature}}$$

The temperature difference ratio is calculated for each element of an external wall. If the result is less than 0.15 the risk of condensation is negligible. If the result is above 0.3 the risk is severe. This approach allows the benefits and risks of improving the thermal performance of the fabric to be assessed at the design stage.

Cold bridge category: T			
Negligible	Moderate	Severe	Unacceptable
< 0.15	0.15–0.2	0.2–0.3	>0.3

(Professor Tadj Oreszczyn, Professor of Energy and Environment, Director of the Bartlett School of Graduate Studies, University College London)

Air pressure testing on an existing building can also be used to find out the number of air changes that occur each hour. This, together with an estimate of the changes after refurbishment work has been carried out can identify future problems.

5.3.2.6 Health warning

In dwellings, condensation is not just a matter of dampness and mould growth. The conditions that allow these to occur lead to an increased risk of asthma in the occupiers. The mechanism seems to be that these conditions are ideal for the breeding cycle of the house dust mite and these mites are directly related to asthma.

5.3.2.7 Attention to detail

If a building relies on mechanical systems to control condensation, it is essential that those systems will work adequately and, as importantly, be used by the occupants. A fan that wakes up the occupants of a dwelling every time a bathroom light is turned on is likely to have its fuse removed. Similarly if a fan with a humidistat runs all the time, whether due to a fault or high humidity level, it is likely to be disabled to prevent noise and stop a perceived waste of electricity.

5.3.2.8 Case study

Soon after the opening of a new office building condensation and mould growth was evident above the windows. The external wall was a cavity wall with a brickwork outer skin. The cavity was fully filled with insulation after construction. Lintels over the windows were a standard folded metal sheet with a bitumastic damp proof membrane acting as a cavity tray. At a late stage in construction the cavity insulation was changed from blown fibre-glass to polystyrene beads.

Opening up above a window showed that the polystyrene beads had reacted to the bitumastic of the damp proof membrane and melted due to the chemical reaction. The lack of any effective insulation in that position created a critical cold spot.

5.3.3 Dampness other than condensation

The cause of any dampness in a building has to be investigated and remedied. If the fabric of a building is damp the use will be compromised.

Damp building material will be colder than adjacent dry material and will be at higher risk of condensation and mould growth. The damp material will also transfer moister to contiguous materials unless separated by an effective damp proof membrane. The result could be damage to fixtures, fitting, furniture and equipment.

5.3.3.1 Investigation

It should be noted that the area that is damp might not be the source of the problem but just the point where the water ends up. Some causes might be simple to identify and simple to fix (e.g. leaking pipe joint needs tightening). Other causes might be complex, difficult to identify and require an extensive knowledge of the construction. Identify the extent of the dampness and the amount of water involved. When no cause is evident keep a record of what occurs. Consider the type of information that will help in the investigation: dates, times, the extent of dampness, weather conditions, internal conditions in the building and external conditions in the surrounding area. Measurements might be useful: the area affected by the dampness, the depth of any standing water, moisture meter readings showing the dampness of materials. These measurements will also help to monitor the success of any remedial action.

Chemical analysis of the water might identify the source. This could show whether the source is piped water, water from the drains, rainwater or ground water. Identify potential sources of dampness:

- Dampness from above (at roof level).
- Dampness from the sides (through the external walls).
- Dampness from below (from the ground or through retaining walls).
- Dampness from within the building (leaking pipes, condensation).

If the source of the water is not apparent, information will be needed about the way the building is constructed.

Dampness from the ground or through retaining walls is a frequent cause of problems in existing buildings. The original construction might not have included damp proof membranes or they could have become ineffective, damaged or bypassed with raised ground levels or water tables.

If damage to a damp proof membrane has occurred in specific areas of a wall, localised repairs could be effective. For more extensive problems other solutions include:

- Removal of the source of dampness (repair leaking drains, reduce ground levels, install land drainage to reduce the water table).
- Insertion of a new damp proof course in a wall.

Physical insertion of a damp proof course might be possible in some situations by cutting through a horizontal joint for the full thickness of a wall and inserting a suitable rigid or flexible damp proof material. This operation would be carried out in sections with a suitable overlap at joints.

If damp is rising up a masonry wall by capillary action, chemicals can be injected to block off the pores. These chemical damp proofing systems might not provide a long-term or cost effective solution and the success or failure depends on a number of factors:

- Correct diagnosis, specification and workmanship.
- Suitable wall construction that will allow the chemicals to block off the pores.
- Replacement of the internal wall finish with a water resistant render to reduce damp and salt penetration.

These systems are less effective when injected into damp walls, as the chemicals cannot displace the moisture in the masonry pores. Voids and loose material in poor quality masonry could prevent the uptake of the chemicals. These problems could prevent the formation of a complete chemical barrier and therefore still permit water to rise up a wall.

5.3.3.2 Case study

A multi storey university hall of residence suffered a defect in the flat roof a few years after opening to residents. The area affected was directly above a vertical service duct carrying water pipes and waste stacks. No defect in the flat roof covering was apparent. The likely cause was thought to be high humidity levels in the duct leading to condensation and deterioration in the boarding that supported the flat roof covering.

No reason for the humidity levels could be identified. The roof was repaired and subsequently failed again after a few years. Further investigation looked for leaking pipe work in the service duct. The timing of the investigation was a key factor in identifying the cause: a leaking pipe to a shower unit that was only evident when the shower was in use. This defect caused a fine spray into the duct leaving no tale-tale drip marks on the walls but raising the humidity to a damaging level.

5.3.4 Thermal insulation

5.3.4.1 Introduction

Most existing buildings are likely to require improvements to thermal insulation either to comply with current legislation, to reduce energy use and running costs or to provide a suitable internal environment.

At the risk of stating the obvious, the main aim of thermal insulation in a building is to stop heat flowing from a warm building to a colder external area. But broader considerations include the total use of resources to provide an environment within the building that is suitable, comfortable and sustainable.

Thermal insulation must be considered alongside other aspects of the built environment including:

- The means of heating and cooling the building.
- Controlling the build up or loss of heat from external and internal sources.
- Insulating equipment and pipe work to reduce heat loss.
- Mechanical and natural ventilation.
- Prevention of condensation.
- Fire protection.
- Acoustic performance.

For existing buildings a balance is needed in the thermal performance of different elements of construction. The introduction of high levels of insulation to external walls whilst retaining poorly performing windows will change the relative temperatures of the elements and could lead to severe condensation problems.

The following steps should be taken when considering thermal insulation:

- Know the building and the thermal performance required.
- Take into account the way the building is used, installation and life cycle costs, timescale, physical and historic limitations.
- Establish target levels for thermal performance.
- Identify appropriate solutions.
- Ensure correct installation and testing.

5.3.4.2 Establishing thermal performance

Thermal energy moves from a high to a low temperature and the larger the temperature difference the faster the energy transfer. The speed of transfer is also affected by the thermal transmittance (U value) of an element of construction. This is a measure of the ability to transmit heat. The lower the U value the better the thermal insulation.

To find the U value of an element, such as a wall, we need to know what it is made of and the way the materials perform. This is established by considering the thermal conductivity (k) of a material and working out the thermal resistance (R) of a specific thickness of that material.

Thermal conductivity (k W/mK) of a material is a measure of the material's ability to transmit heat and is expressed in watts per square metre of surface area for a temperature gradient of one kelvin (K) for a one metre thickness of material. The lower the thermal conductivity, the better the insulating properties of the material.

Thermal resistance (R m²K/W) of a specific thickness of a material can then be calculated from R = t/k where t = the thickness of the material in metres. This is a measure of resistance to the passage of heat of that thickness of material.

The performance of the complete element of construction is the thermal transmittance (U value) of all the materials that make up that element and is found by taking the reciprocal of the sum of the thermal resistances of all the parts of the construction. Thus U=1/R W/m² K.

5.3.4.3 Establishing target levels of thermal performance

Over 45% of the UK carbon dioxide (CO_2) emissions are produced by the energy consumption of buildings, including the operation of associated building services. The pressure to reduce the energy used is intensive and increasing.

The loss of heat through the walls, floor and roof of a building is determined by the U value of the construction. The U value is the thermal transmittance of the construction measured in W/m²K (watts per metre of area, degrees Kelvin). The U value is a property of the whole construction, the materials, any spaces in the construction and the surfaces of the elements. As it is a measure of the ability of the construction to transmit heat the lower the U value the better the thermal insulation.

U values for most types of construction are available or can be calculated if the thermal resistance of each material, space and the surface is known. See Table 5.8 for examples.

5.3.4.4 An outline of Part L of the Building Regulations

For most projects, Part L of the Building Regulations will determine the minimum targets for thermal insulation. A major revision to Part L came into force in April 2006. Parts L1A and L1B apply respectively to new and existing dwellings; Parts L2A and L2B apply respectively to new and existing buildings other than dwellings. The means of complying with the Regulations requires a good working knowledge of the subject.

Table 5.8 Typical U values of some constructions W/m²K.

Element	U Value
Solid 220 mm brick wall plastered internally	3.30
Cavity wall with 100 mm brickwork, 100 mm cavity fully filled with insulation, 100 mm dense blockwork inner leaf, plasterboard internally	0.37
As above with lightweight blockwork inner leaf	0.26
Wood window with single glazing	4.30
Wood window with double glazing	2.50

Part L requires reasonable provision for the conservation of fuel and power in buildings by:

- Limiting heat gains and losses through thermal elements and other parts of the building fabric and from pipes, ducts and vessels used for space heating, cooling and hot water services.
- Providing and commissioning energy efficient fixed building services (such as boilers) with effective controls.
- Providing information to the owner about fixed services and their maintenance requirements so that fuel and power is not wasted.

Compliance with the building regulations is a legal requirement and approval is required before projects are carried out.

Some work is exempt from some requirements of the building regulations. There are different requirements for new and existing buildings and there can be exemptions or reduced targets, in some situations, and for historic buildings.

Complying with Part L

The way of complying with Part L is different for new and existing buildings.

For new buildings in the current building regulations there is now only one approach to show compliance with the energy efficiency requirements. This is based on five criteria:

- The annual CO_2 emission rate of the completed building, as calculated using SAP 2005 (Standard Assessment Procedure), must not exceed set targets.
- Building fabric and service performance specification are within reasonable limits.

- Solar shading and other measures to limit risk of summer overheating are reasonable.
- Fabric insulation and airtightness standards are met.
- Satisfactory information is provided to the occupiers to achieve energy efficiency in use.

For existing buildings an elemental approach is used. This sets out target U values for elements of construction. The targets, and whether any change is needed to the building, depend on a number of factors. The following information is intended to give an indication of the requirements of the regulations but reference to the actual documents is essential.

If the thermal elements are newly constructed, such as for an extension, or they are a replacement, reasonable provision for thermal elements in W/m^2K (U value) is as follows (see Table 5.9):

Table 5.9 New and replacement thermal elements.

Element	Standard for new thermal elements in an extension	Standard for replacement thermal elements in an existing dwelling
Wall	0.30	0.35
Pitched roof	0.16	0.16
Flat roof	0.20	0.25
Floors	0.22	0.25

If there is a renovation of a thermal element, the 'improved value' shown below should be achieved. If such an upgrade is not technically or functionally feasible or would not achieve a simple payback within 15 years, an upgrade to a suitable payback standard should be set.

If there is a retained thermal element, together with a material change of use of the building, a reasonable provision is required to upgrade those elements that have a U value worse that the 'threshold values' shown in Table 5.10. The aim is to achieve the 'improved values' subject to the 15-year simple payback test.

5.3.4.6 How to insulate your building

Thermal insulation is only one aspect of the way the complete fabric of the building performs. It cannot be treated in isolation from the way the building is used, the mass of the fabric, heating systems, ventilation, condensation, vapour movement, fire performance and cost effectiveness.

Building materials such as lightweight block work and timber have useful thermal insulation properties alongside their other functions. But to achieve the required levels of thermal performance in a building, purpose made products are needed such as

Table 5.10 Upgrading retained thermal elements W/m²K (U value).

Element	Threshold value	Improved value
Wall	0.70	0.55
Pitched roof	0.35	0.16
Flat roof	0.35	0.20
Floors	0.70	0.25

mineral fibre quilt, expanded polystyrene and urethane, reflective foils, vacuums and gases. The nature of these products means they are concealed within the construction behind materials that provide suitable surface finishes or weather protection.

Increased insulation can be applied to walls, roofs and floors. The decision on what steps to take must consider the risks involved in altering the performance of a building's fabric. This is dealt with in detail in the Building Research Establishment document BR262 *Thermal Insulation: avoiding risks*. The 2002 edition of the document remains relevant. It gives comprehensive data with accompanying diagrams on constructional details for the overall building, floors, walls, windows and roofs. Guidance is also given on how to determine the suitability of wall insulation for different exposure conditions in the UK.

The Department for Communities and Local Government (DCLG) currently publish a set of details to help the construction industry achieve the energy efficiency requirements of Part L of the Building Regulations. These are known as Accredited Construction Details (ACDs). The details and introductory section of the document focus on the issues of insulation continuity (minimising cold bridging) and airtightness. They are not intended to provide any detailed guidance on other performance aspects such as vapour control, ventilation, etc which must also be considered by the design and construction team.

For advice on energy the Energy Saving Trust is a non-profit organisation that promotes energy saving and is funded by the government and the private sector. It was set up after the 1992 Rio Earth Summit with two main goals:

- To achieve the sustainable use of energy.
- To cut carbon dioxide emissions, one of the key contributors to climate change.

To achieve these goals, they work with households, businesses and the public sector:

- Encouraging a more efficient use of energy.
- Stimulating the demand and supply of cleaner fuelled vehicles.
- Promoting the use of small-scale renewable energy sources, such as solar and wind.

Further information is available on the Trust's website (see Appendix, Section A2.3).

Establishing how heat is lost from a building will help to identify the most effective approach to improving insulation. For a poorly insulated house the typical heat loss through the fabric are: 25% through the roof, 15% through the floor, 35% through the wall fabric and about 25% through windows, doors and draughts.

5.3.4.7 Walls

To improve the insulation of an existing wall the options are:

- Add insulation internally.
- Add insulation externally.
- Add insulation within the depth of the construction.
- Improving windows and doors.
- A mixture of the above.

The building type and construction will influence or determine the approach. If a building is occupied, internal insulation might be too disruptive. Changing the appearance of a structure with external insulation might not be acceptable. Solid external walls would preclude adding insulation in the wall depth. Historic considerations might prevent alterations to windows and doors.

General items for walls

Carry out investigation and remedial work to ensure the wall is suitable for its intended purpose including:

- Structural suitability.
- Providing adequate weather protection.
- Effective protection from rising damp.
- Internal or cavity insulation will cause the outer fabric to be colder and more susceptible to damp, frost and sulfate damage.

Consider removing internal and external wall finishes if:

- They are defective or unsound or not suitable for the intended construction.
- They are likely to absorb and retain moisture.
- The finishes act as a vapour check in the wrong position. Vapour control layers (to prevent vapour entering into the wall) should be on the warm side of the construction and the wall should become more vapour permeable towards the outside of the building.

Internal wall insulation

Typical construction and materials:

- Insulation is added to the inner face of the wall and provided with a suitable surface finish. Skirting, architraves, door and window reveals are adapted to suit.

- Plasterboard with an insulating layer fixed to the inner wall face by mechanical or adhesive means, with an integral vapour control layer or surface applied vapour check.
- Studwork wall constructed on the inner face of the wall with mineral fibre quilt or expanded foam insulation. Faced with plasterboard with an integral vapour control layer or surface applied vapour check. This would also apply to upgrading an existing timber framed inner wall lining.

Key items are:

- Vapour control layer on the warm side of the wall.
- Seal gaps at floor, ceilings and reveals to prevent vapour entering the construction.
- Cold bridging at floor and roof level and any structural projection from the building such as balconies.
- Cold bridging at services where they penetrate the wall.
- Fire spread through the insulation material and cavities.
- Overheating of electrical cables within insulated wall.
- Provide users with information about the wall construction.

External wall insulation

Typical construction and materials:

- Insulation is added to the external face of the wall and provided with a suitable surface finish. Fixing of insulation is by mechanical or adhesive techniques.
- Expanded polystyrene fixed to the wall and finished with a render to provide weather resistance and an acceptable appearance.
- Rigid sheets of mineral fibre insulation fixed to the wall and finished as above.
- Finished with a traditional sand cement renders on a suitable supporting background or reinforcement.
- Finished with a thin coat polymer modified render applied with mesh reinforcement in the first layer and a decorative second coat.
- Finished with suitable tiles, sheets or boards.

Key items are:

- Suitable vapour control layers and vapour permeability of the insulation system.
- Impact damage and cracking of the render. Areas at lower level might need a masonry wall to provide protection.
- Fire propagation due to cavities and combustible materials.
- Cold bridging at floor and roof level and any structural projection from the building such as balconies.

- Limited or no improvement in insulation at reveals if existing window size is retained.
- Reductions in insulation thickness at rain water pipes and other external services leading to colder areas on internal walls.
- Cold bridging at services where they penetrate the wall.
- Avoid blocking air vents and flues.

Insulation within the depth of the wall

Typical construction and materials:

- Insulation is added within the depth of the construction by filling a cavity with insulation. Holes are drilled in the external leaf of a cavity wall and insulation blown or injected into the cavity.
- Mineral fibres blown into the cavity.
- Expanded polystyrene beads blown into the cavity.
- Injection of expanding polyurethane foam into the cavity.

Key items:

- Suitability of the cavity to be insulated to prevent insulation escaping into other parts of the construction.
- Vapour control layer on the warm side of the wall.
- Compatibility with other material. Polystyrene beads could be damaged by chemical reaction with bitumastic and other materials in damp proof membranes.
- Cold bridging at floor and roof level and any structural projection from the building such as balconies.
- Colder and wetter conditions in the cavity could cause deterioration in wall ties.
- Unsuitable wall ties could allow water across the cavity.
- Mortar dropping on wall ties could create cold bridges across the cavity.
- Limited or no improvement in insulation at reveals and lintels.
- Reductions in cavity width at structural columns and piers and reduction in the insulation value leading to cold areas on internal walls.
- Cold bridging at services where they penetrate the wall.
- Seal at floor, ceilings and reveals to prevent vapour entering the construction.
- Fire spread through the insulation material and cavities.
- Problems with fumes

Improved windows and doors

A double glazed timber window has a U value that is around 40% better than single glazing. The heat lost through windows and doors is very high compared to well-insulated walls so any improvement is of great benefit. Older windows and doors are less likely to have draught seals so the air changes in the building, with the consequent loss of heat, will be high. The following should be considered:

- Replacement of windows and doors to achieve the thermal performance required.
- The addition of secondary glazing.
- Improvement to or addition of draught strips.

If the windows and doors have better draught seals the lower rate of air changes within the building will reduce energy use but could lead to condensation problems.

The overall performance of any replacement window or door should be considered. A double glazed window with a cold metal frame will have the potential for severe condensation on the frame. If metal frames are used they should be thermally broken so the external temperature of the metal is not conducted to the inner part of the frame.

The coldest part of an external wall will normally be the centre of the glass of a single glazed window. If this is change to a double glazed window; the coldest part is likely to be the wall above the window. A balanced approach is needed to avoid condensation problems being created or made worse.

For a single glazed window the glass provides very little insulation. By trapping air between two panes of glass, double-glazing creates an insulating barrier that reduces heat loss, and reduces noise and condensation. The efficiency can be improved by the use of vacuums and inert gases. Further layers of glass and gas give further improvements. Triple glazing is used in colder climates and can achieve very low thermal transmittance values. The U values achievable for high performance triple glazed windows using different glass and details are:

- Standard Glass 2.0 W/m^2K
- Low E Glass 1.6 W/m^2K
- Low E Glass and Argon (Basic Specification) 1.4 W/m^2K
- Low E Glass, Argon and using a warm edge detail 1.2 W/m^2K

Low E glass stands for low emissivity glass where one side of the glass has a transparent metal coating that increases the energy efficiency of windows by reducing the transfer of heat or cold through the glass.

The sun's short wave energy passes through a window and is absorbed by carpet, furniture, etc. and is transformed into long wave radiation. The Low E coating reflects the radiant room-side heat back into the building. In summer the coating reduces the amount of the sun's short-wave radiation entering the building.

High levels of insulation and around 13% light transmission can be achieved with recently developed building products. A translucent insulation is faced with translucent fibreglass sheets and offer U values from 2.7 to 0.56 W/m^2K. The core insulation material, translucent aerogel, was developed for scientific purposes. At present the high cost is limiting its application in buildings.

5.3.4.8 Roofs

To improve the insulation of an existing roof the typical options are:

- Add insulation below ceiling level.
- Add insulation within the structural depth of the roof.
- Add insulation under the weatherproof surface of the roof.
- Add insulation above the weatherproof surface of the roof.
- In some circumstances a mixture of the above could be used.

General items for roofs

Carry out investigation and remedial work to ensure the roof is suitable for its intended purpose including:

- Structural suitability.
- Providing adequate weather protection.
- Suitable for increased thermal insulation performance.

The following should be considered:

- Increased insulation will result in lower temperatures on the cold side of the insulation.
- The insulation of tanks and pipes need to be considered.
- The risk of condensation in materials and voids will change and could increase.
- Ventilate roof spaces and voids where there is a risk of condensation.
- Electric cables and equipment need to be suitable for covering with insulation.
- Penetrations through ceilings and insulation need to be considered. Items such as recessed lighting could penetrate vapour control layers, insulation and fire protective layers.

Ventilation to roof voids

There are requirements in the building regulations for the ventilation of roof spaces and voids where there is a risk of condensation and these have to be considered at an early stage of the design. Reference is made to the Building Research Establishment document BR262 *Thermal Insulation: avoiding risks* for guidance on the requirements.

If moisture passes through the ceiling there is a risk of condensation occurring within the insulation and on cold surfaces in the roof space. Voids above insulation will normally require ventilation to the outside air to reduce condensation risks. The required ventilation can alter the details and the overall depth of the roof construction. The current recommendations are:

- Above the insulation in pitched roofs provide the equivalent of a continuous ventilation gap of 5 mm at the ridge, and 25 mm at the eaves for a pitch of 15 degrees or below and 10 mm at the eaves for higher pitches.

- Above insulation in flat roofs a continuous cross ventilation path of 50 mm is required for a span of up to 5 m, with eaves ventilation of 25 mm. For spans from 5 to 10 m the figures are 60 and 30 mm respectively or 0.6% of the roof plan area, taking whichever is the greater.

If the correct ventilation cannot be achieved other solutions will be required for insulating the roof.

Insulation below ceiling level of a flat or pitched roof

Typical construction and materials:

- Plasterboard with an insulating layer fixed to the underside of the ceiling, or ceiling structure, by mechanical or adhesive means. With an integral vapour control layer or surface applied vapour check.
- Timber or metal battens fixed to the underside of the ceiling, or ceiling structure, by mechanical means with mineral fibre quilt or expanded foam insulation. Faced with plasterboard with an integral vapour control layer or surface applied vapour check.

Key items:

- Voids above insulation will normally require ventilation to the outside air to reduce condensation risks. For an existing flat roof this might not be possible and an alternative approach would be required.
- Maintain ventilation paths at eaves and junctions with other elements.
- Install vapour control layer on the warm side of the insulation.
- Details at junctions to prevent cold bridging.
- Maintain continuity of vapour control layers, insulation and fire protective layers where recessed lights and other penetrations occur. If necessary box out above equipment with suitable construction.
- Seal around edges and at any penetration through the ceiling.
- Fire risks in voids and from any combustible materials.

Insulation within the structural depth of a flat roof or at ceiling level in a pitched roof

Typical construction and materials:

- Mineral quilt or expanded foam installed above ceiling level between structural members.
- For a flat roof the insulation would be within the structural depth.
- For a pitched roof the insulation should also go between and over the ceiling joists to reduce cold bridging.

Key items:

- Voids above insulation will normally require ventilation to the outside air to reduce condensation risks. For an existing flat roof this might not be possible.

- Maintain ventilation paths at eaves and junctions with other elements.
- Install vapour control layer on the warm side of the insulation.
- Details at junctions to prevent cold bridging.
- Maintain continuity of vapour control layers, insulation and fire protective layers where recessed lights and other penetrations occur.
- Seal around edges and at any penetration through the ceiling.
- Fire risks from any combustible materials.
- Insulate pipes and tanks.

Insulation under the weatherproof surface: warm roof construction for flat roofs

For a warm roof the insulation is placed above the roof deck but below the weatherproof membrane. The insulation is bedded on a continuous high performance vapour control layer. This prevents vapour entering the insulation and reduces the risk of condensation in the construction. No ventilation is required for this form of roof.

Typical construction and materials:

- The roof deck could be timber, concrete or metal sheeting.
- Rigid mineral fibre slabs or expanded foam board insulation.
- Roofing felt, asphalt, metal decking or proprietary membranes as weatherproof layers.

Key items:

- Correct selection and use of material to provide a workable solution.
- Install vapour control layer on the warm side of the roof if necessary to reduce vapour entering any unheated voids.
- Details at junctions to prevent cold bridging.
- Tight fitting of insulation to prevent cold bridging.

Insulation above the outer surface of the roof: warm roof construction for flat roofs

This is a variation on the above construction where the insulation is placed above the waterproof membrane.

Typical construction and materials:

- The roof deck could be timber or concrete.
- Expanded foam insulation.
- Protection of the insulation from ultraviolet and physical damage.
- Means of stopping lightweight insulation being blown away

Key items:

- Correct selection of materials and an allowance for rain cooling of the insulation.
- Protection of the insulation can be by integral screed surface, paving slabs, or gravel.

- Provide pedestrian routes if access is needed.
- Use correct details at perimeters, rainwater gullies, and roof penetration.

5.3.4.9 Floors

Heat is lost through floors and at the junction of a floor and wall. This applies to floors at ground level and those above unheated spaces. For most projects Part L of the Building Regulations will determine the minimum targets for the thermal insulation of floors for new construction, replacement and upgrading of thermal elements.

Where meeting the targets would create significant problems in relation to adjoining floor levels a lesser provision may be permitted.

The U value for a ground floor depends not only on its construction but also on size, shape, soil type and edge insulation.

To improve the insulation of an existing floor the options are:

- Add insulation below the floor.
- Add insulation above the floor.
- Add insulation within the depth of the construction.
- Add insulation to the edge of the floor.
- A mixture of the above.

The building type and construction will influence or determine the approach. Adding insulation below an existing concrete ground floor would not be feasible. Raising the floor level by insulating above a floor would affect room heights and staircases. Insulating within the depth of a construction would apply to specific types of construction. Replacement of floors might be required in some situations.

General items

Carry out investigation and remedial work to ensure the floor is suitable for its intended purpose including:

- Structural suitability.
- Effective protection from rising damp.
- Condition of screeds and finishes to perform in new conditions.
- Suitability for increased thermal insulation performance.

The following should be considered:

- Increased insulation will result in lower temperatures on the cold side of the insulation.
- The insulation of incoming services needs to be considered.
- The risk of condensation in materials and voids will change and could increase.
- Electric cables and equipment need to be suitable for covering with insulation.

Insulation below a ground bearing concrete floor

Typical construction and materials:

- this applies to ground floors
- Insulation laid on hardcore and used as a base for casting the concrete slab.
- Expanded foam boards or mineral fibre.

Key items:

- Insulation must have suitable compressive strength, moisture resistance and resistance to any chemicals in the soil.
- Lay on adequate base to avoid deformation when concrete slab is cast.
- Use damp proof membranes in suitable locations in the floor and at edges.
- Avoid damage during construction.
- Avoid condensation from thermal bridging.
- Ensure continuity of floor and wall insulation at the edges to prevent cold bridging.

Insulation below a suspended concrete floor.

Typical construction and materials:

- This applies to ground floors where access is available underneath and to floors above unheated spaces.
- Plasterboard or similar with an insulating layer fixed to the soffit of the structure, by mechanical or adhesive means. With an integral vapour control layer or surface applied vapour check.
- Timber or metal battens fixed to the underside of the soffit by mechanical means with mineral fibre quilt or expanded foam insulation. Faced with plasterboard or similar with an integral vapour control layer or surface applied vapour check.

Key items:

- Plasterboard or similar must be suitable for moisture conditions.
- Avoid condensation from thermal bridging.
- Ensure continuity of floor and wall insulation at the edges to prevent cold bridging.
- Seal around services and penetrations.

Insulation above the floor

Typical construction and materials:

- This applies to ground floors and floors above unheated spaces.
- Insulation laid on a concrete slab, on precast concrete flooring or timber construction.
- Expanded foam boards or mineral fibre.
- Surface above the insulation could be a screed or a timber based flooring.

Key items:

- Insulation must have suitable compressive strength and moisture resistance.
- Use damp proof membranes and vapour control layers in suitable locations in the floor and at edges.
- Avoid damage during construction.
- Avoid condensation from thermal bridging.
- Ensure continuity of floor and wall insulation at the edges to prevent cold bridging.
- Integrate services where required.

Insulation within the depth of the construction

Typical construction and materials:

- Where a timber or metal beam structure is used the void could be suitable for insulating. This applies to ground floors and floors above unheated spaces.
- Expanded foam boards or mineral fibre insulation fixed between the structural members.
- To prevent cold bridging, especially with a steel structure, insulation under the structural members might be needed.

Key items:

- Fire risk if combustible materials used and cavities and voids created.
- Ventilation of the space under a ground level floor is necessary to prevent dampness and condensation.
- Sub floor services will be colder and could require protection from frost.
- Seal gaps in construction to prevent cold air entering the building.
- Avoid condensation from thermal bridging.
- Ensure continuity of floor and wall insulation at the edges to prevent cold bridging.
- For a ground level floor a vapour control layer is not needed as any condensation that forms will be removed by ventilation.
- For a floor above an unheated space a vapour control layer might be needed if moisture is present and ventilation is poor.

Insulation to the edge of a concrete ground floor.

Typical construction and materials:

- Edge insulation could provide adequate performance for some floors.
- For new construction expanded foam boards or mineral fibre can be used vertically (against the wall) or horizontally (under the floor structure) and in wall cavities. Blockwork with a good thermal performance can also be used to provide an insulated wall and avoid cold bridges at the junction with the floor.

- For remedial work vertical insulation can be applied to the outer face of the wall above and below ground.

Key items:

- Insulation must have suitable compressive strength, moisture resistance and resistance to any chemicals in the soil.
- Use damp proof membranes in suitable locations in the floor and at edges.
- Avoid damage during construction.
- Avoid condensation from thermal bridging.
- Ensure continuity of floor and wall insulation at the edges to prevent cold bridging.
- Insulation on the external wall face requires suitable protection from damage.

5.3.5 Sound insulation

5.3.5.1 Introduction

Achieving a suitable level of sound insulation in a refurbished building can be a challenging task. The challenges will only be met if consideration is given to all relevant factors at the earliest stages of the design process. The following steps should be taken:

- Know the building and the criteria for the work. Consider cost, timescale, physical and historic limitation etc.
- Establish the target levels of sound insulation in different areas of the building.
- Identify effective technical solutions to achieve those levels.
- Ensure the building work is carried out satisfactorily.
- Test the results where necessary.

5.3.5.2 General process

Assess what the options are for the building type you are dealing with. For many types of sound insulation there will be the need for the addition of new layers of construction. This might require ceiling height to be lowered, floor heights to be raised and wall thicknesses increased. If existing staircases are to remain, raising floor heights might not be an option. Most forms of sound insulation are likely to increase the weight of the construction. Make sure any additional loads can be carried by the existing elements of construction. In historic buildings preservation of ceilings, flooring and wall finishes could limit the possible range of solutions. Other requirements, such as improvements in fire protection or thermal insulation also need to be taken into account when identifying technical solutions.

Part E of the Building Regulations sets out the legal minimum levels of sound insulation only for residential accommodation and schools. For other building types the activities, owner's requirements, the users or the location of the building will determine the levels of insulation that are needed.

In some situations conditions attached to planning approvals will state the level of sound insulation required. This could apply where residential accommodation is close to noisy road or railways. Sound is a two way process. If noise generation within a building is excessive, steps need to be taken to reduce the escape of sound. Various types of legislation would determine the requirements in this area.

Where it is necessary to increase the level of sound insulation, satisfactory results will only be achieved if the design and construction is well thought through and carried out correctly. All aspects of the design and construction need to be considered. The effectiveness of a well specified and constructed sound-resisting wall could be ruined by a small hole drilled though it.

A thorough understanding of the construction of the building and the way sound is transmitted through the structure is necessary. The aim of sound insulation is to reduce the sound that passes from one area to another. Sound can be transmitted in the following ways:

- Direct transmission. An example of this would be the sound from one area passing directly through a wall into the next room.
- Indirect transmission (also known as flanking transmission). Examples include open windows, pipe routes and ducts and transmission through walls, floors and ceilings that flank the two areas.
- Sound can be airborne between two areas (e.g. someone talking in one room being heard in the adjacent room) or impact sound (e.g. heavy footsteps on a floor).

Where there is a sound transmission problem it is essential to identify the weak points in the construction.

5.3.5.3 *Establishing the levels of sound insulation*

The level of sound is measured in decibels (dB) by using a sound meter. Sound insulation is the way that sound transmission is reduced. The effectiveness of sound insulation varies with the frequency of a sound wave; therefore a limited range of frequencies that make up everyday sounds is used for measurement.

Typical sound insulation levels for different activities are:

25 dB	normal speech can be heard easily
30 dB	loud speech can be heard easily
35 dB	loud speech can be distinguished under normal conditions
40 dB	loud speech can be heard but not clearly distinguished
45 dB	loud speech can be heard faintly
50 dB	shouting is difficult to hear

These figures are the sound insulation of building elements as measured in the laboratory and are the Weighted Sound Reduction Index (Rw). This single number measurement describes the overall acoustic performance of a part of a building and shows the performance that is achieved or required of a floor, wall or roof. These sound insulation rating methods are defined in BS EN 717-1: 1997 (airborne) and BS EN 717-2:1997 (impact).

The Building Regulations set out the performance standards for existing buildings that are being converted to residential use as follows:

- Walls separating dwellings: 43 dB as a minimum airborne sound insulation.
- Floors and stairs separating dwellings: 43 dB as a minimum airborne sound insulation value and 64 dB as a maximum impact sound insulation value.

The way these values are calculated is set out in the Building Regulations and requirements are also given for purpose-built houses and flats and for historic buildings. Testing the level of sound insulation achieved in some buildings is required for compliance with some sections of the Building Regulations.

Further information on sound insulation is available from a number of sources including the British Gypsum White Book that also sets out the theory alongside possible solutions. The British Gypsum website provides further details.

Establishing the solutions for new construction

Looking at different types of new construction and the way sound insulation is achieved provides useful background in understanding how to deal with sound.

For separating walls, the typical types of construction and the way they achieve their sound insulation are as follows:

- Solid masonry wall plastered both sides. The resistance to airborne sound is achieved mainly by the mass of the wall (i.e. kg/m^2).
- Cavity masonry plastered both sides. The resistance to airborne sound depends on the mass of the wall (i.e. kg/m^2) and the degree of isolation between the masonry. To be most effective the isolation has to limit the connection between the two leaves of masonry. Avoid solid connections across the cavity including floor structure, beams, pipes etc. Wall ties should be twisted wire and not rigid metal.
- Solid masonry wall between independent panels. The resistance to airborne sound depends partly on the mass of the wall (i.e. kg/m^2) and partly on the isolation and mass of the independent panels.
- Framed wall with absorbent material. The resistance to airborne sound depends on the mass of the leaves (i.e. kg/m^2), the isolation of their supporting frames and the sound absorption contained within the cavity.

For any wall construction to be effective in providing sound insulation, care is needed at the junction between the separating walls and other elements such as floors, roofs, external an internal walls.

For separating floors the typical types of new construction and the ways they achieve their sound insulation are as follows:

- Concrete slab with a ceiling and soft floor covering. The resistance to airborne sound is achieved mainly by the mass of the concrete and the mass of the ceiling construction (i.e. kg/m^2). The soft floor covering reduces the impact sound at source.
- Concrete slab with a ceiling and a floating floor. The resistance to airborne and impact sound depends on the mass of the concrete slab and the mass and isolation of the floating layer and the ceiling. The floating floor is typically a sand cement screed on a resilient layer on top of the slab and this reduces the impact sound at source. The floating floor could also be timber on a resilient layer.
- Timber joists supporting a platform floor with independent ceiling. This provides a floor supported on timber joists and a separate ceiling supported on its own timber structure. The resistance to airborne and impact sound depends on the structural floor base and the isolation of the platform floor and the ceiling. The platform floor reduces impact sound at source.

A series of Robust Standard Details that are compliant with regulation E of the Building Regulations can be accessed on the Robust Details website (www.robust-details.com).

5.3.5.4 *Establishing the solutions for existing construction*

For existing buildings the same technical solutions could be applied if the construction is similar to those described above. For the category of building dealt with in the Building Regulations Part E sets out the requirements for the minimum mass of flooring in existing buildings, requirements for increasing mass and upgrading floors walls and ceilings. Guidance is given on the possible extensive work required to reduce flanking transmission. Examples of typical solutions to upgrading include the following:

- Independent wall panels with absorbent materials. A new stud wall is constructed close to but independent of the existing wall. The resistance to airborne sound depends on the form of the existing construction, the mass of the panels, the isolation of the panel and the absorbent materials.
- Floor treatment using independent ceiling with absorbent material. A new ceiling with its own structure is constructed underneath but separated from the existing floor. The resistance to airborne and impact sound depends on the mass of the existing floor and the new ceiling, the isolation of the ceiling, the absorbent material and the sealing of the construction to prevent sound passage.
- Floor treatment using a platform floor with absorbent material. A resilient material is laid on the existing floor and supports new floating floor surface. The resistance to airborne and impact sound depends on the total mass, the effectiveness of the resilient layer and the performance of the absorbent layer.

Proprietary systems are available for many applications. Some will allow the upgrading of an existing timber floor with little or no change in the floor or ceiling level. Systems have the potential advantage of tested products to achieve certain levels of insulation, manufacturer's information and technical advice.

Where fully independent walls and ceiling cannot be constructed due to limited space or structural limitations, there are proprietary connectors that allow the connection of different elements but limit the amount of sound that will be transmitted through the connector. (See also the Robust Details website: www.robustdetails.com.)

5.3.5.6 Ensuring the building work is carried out satisfactorily

To achieve the expected levels of sound reduction the building work must be carried out correctly. The following are key items to consider:

- If the construction is to be independent of other structure, ensure that it is.
- Ensure any independent panels and frames are isolated from the existing wall.
- Details, such as skirtings and architraves, must not be in contact with an independent floor construction.
- Seal the perimeter of the panel to prevent holes and routes for sound.
- For absorption layers avoid tight compression of the material.
- Specify and use the correct materials. The density of boards and resilient layers is important. The use of the correct type of wall ties is important.
- Allow for movement in the finishes and structure.
- Take care around any penetrations in the construction.
- To be certain, test the final results at least in typical areas. If they fail, carry out remedial work and test again. Then test similar areas to see if they fail. The alternative to this systematic and thorough approach is the unsatisfactory solution of waiting for users to complain and then having to carry out the test and the remedial work!
- Where resilient layers support a floating screed, make sure the joints are tightly butted to avoid screed bridging the resilient layer.
- Follow manufacturer's instructions and guidance.
- Pay attention to hidden parts of construction such as those above partitions and suspended ceilings.
- Fill holes in construction with the appropriate materials. Masonry walls must have all the joints fully filled to prevent sound paths. Gaps between independent constructions should be filled with flexible sealant.
- Doors and windows will be weak points for sound insulation. Specify and detail to suit the requirements of the building.
- Make sure the owner and the users are aware of the technical requirements of the building. Written information must be made available to owners

and those who will maintain and alter a building so they are aware of the construction and technical requirements.

- The sound absorption materials also act as thermal insulation. Make sure any electrical cables and equipment are suitable for insulated situations.
- Make sure that the requirements for sound protection from external sources, such as roads, is known and allowed for. This could affect external wall construction, the type of window used, the junction of the window to the wall and the provision of insulated trickle vents separate from the window. Double glazing should also be considered with suitable gaps between glass panels to achieve required sound reduction.

5.3.5.7 Conclusions

Sound insulation is a difficult subject to deal with. Good results can be achieved by a thorough understanding of the building, the insulation requirements, the available solutions and the process of achieving those solutions. A small failure in one part of the process could undermine a project.

5.3.6 Fire (including Fire engineering)

5.3.6.1 Historic background

For practitioners wishing to learn more about fire and fire precautions in early buildings an excellent starting point is Thomas Swailes paper *19th century fireproof buildings, their strength and robustness* (2003). This concludes with a list of more than 50 references. Swailes deals with the so-called fireproof buildings used in Britain from the end of the 18th century after disastrous fires in very large timber-framed factories. He goes on to say that reliance was made of the use of incombustible materials: brick arch or filler joist slab floors supported on iron beams and columns. Such construction cannot easily be justified by calculation but their composite action may be verified by load test. Enhancement, where necessary, may be effected by the provision of new structural topping. Further background information is provided by synopses of collapses of a number of structures including:

- Gough's mill, Salford, 1824.
- Radcliffe's cotton mill, Oldham, 1844.
- Prison building, Northleach, 1844.
- Boyd's flax mill, Belfast, 1851.
- Office building, Gracechurch Street, London, 1851.
- Malt barn, Edinburgh, 1857.
- Flax spinning mill, Aberdeen 1862.
- Sugar refining building, Leith, 1865.
- King's College dining hall, London, 1869.
- Ropeworks, Greenock, *c.*1997.

- Mill, Hull, 2000.
- Roof terrace, London, 2002.
- McConnel and Kennedy's mills, Manchester, 1997–2002.

Although none of these collapses occurred due to fire, an examination of these reports gives a good insight into the construction details in use at those times.

The Building Regulations requirement for fire (which is designed primarily to ensure safety of life) is stated as: 'The building shall be designed and constructed so that, in the event of fire, its stability will be maintained for a reasonable period'. In England and Wales, an Approved Document (AP) approach is used to satisfy this requirement. The relevant document in this case is AP B in which a combination of building occupancy and height determine the required fire resistance. This can vary from 15 to 120 minutes but the dominant period by far is 60 minutes.

As with many topics in the construction area, the London County Council (LCC) in the London Building Acts (1930–1939) set out the aims and objectives of fire resistance with a clarity that has probably not been surpassed. The statement reads:

The purpose of this part of these by-laws is to minimise the risk of the spread of fire between adjoining buildings by a stable and durable form of construction to prevent the untimely collapse of buildings in the event of fire and to minimise the risk of the spread of fire between specified parts of buildings.'

Although the LCC By-laws (and their successors with the Greater London Council (GLC)) were only applicable in inner London, their requirements were frequently applied in outer London and other parts of England.

In the last thirty years or so, the approach to fire resistance in structures has changed from a prescriptive classification system to one where a specific analysis is made of each individual case. In the first, a building or part of a building was granted a classification in terms of hours of resistance related to the class of structure. Fire resistance was to a large extent based on the performance of elements subjected to standard fire tests to BS476. In the latter, using fire engineering techniques, a range of parameters are considered.

Practitioners involved in refurbishment work must be careful not to temporarily jeopardise the integrity of fire protection. On a recent London redevelopment, cast iron columns that had been previously fire-protected were left unprotected for the duration of reconstruction work, putting at risk (for several months) upper stories of a four-storey block of flats.

5.3.6.2 Fire Protection Engineering

- Active fire protection including fire detection, evacuation procedures, alarm systems, sprinklers and other automatic fire fighting systems.
- Passive fire protection dealing with the design of a building for adequate load bearing resistance and for limiting fire spread under fire conditions. This discipline is generally categorized as structural fire engineering (SFE).

SFE can be divided into three levels of complexity:

- Simple procedures as given in most conventional building codes.
- Calculation of structural fire resistance based on the empirical or theoretical relationships.
- The assessment of three basic aspects comprising the likely fire behaviour, heat transfer to the structure and the structural response.

In recent times a series of whole building fire tests (on concrete, steel and timber framed multi-storey buildings) at the BRE facility at Cardington have influenced the way engineers look at these matters.

The University of Manchester, under Professor Colin Bailey, has set up a website to act as a One Stop Shop in Structural Fire Engineering (www.mace.manchester. ac.uk/project/research/structures/strucfire). For legal requirements see Chapter 6 in this book.

5.3.6.3 Fire in concrete structures

Well designed and constructed reinforced concrete has an inherent fire resistance. BS 8110-1 states that:

> a structure or element required to have fire resistance should be designed to possess an appropriate degree of resistance to flame penetration, heat transmission and collapse.

BS 8110-2 gives recommendations for cover to reinforcement based on [element shape and] constituents allowing benefit for additional protection such as gypsum plaster. It also states that the fire resistance of whole concrete structures may be greater than that ascribed to its individual elements. Reinforcement of cold worked steel shows a rapid decrease in strength after 300°C although in well designed and constructed concrete this should be adequately protected from fire by cover to the reinforcement.

As a guide to practice in the days when permissible stress philosophy for the design of reinforced concrete reference to Table 5.11 (reproduced from IStructE *RC Permissible stress recommendations* Table 25) may be used in assessing existing structures designed and built before the 1990s and possibly some beyond that date.

5.3.6.4 Fire in metal structures

British Steel Swindon Technology Centre has produced a helpful guide on the effects of fire in iron and steel structures. Further information on this can be found in a review by Barnfield and Porter (1984).

Hot finished carbon steel begins to lose strength at temperatures above 300°C and reduces in strength at a steady rate up to 800°C. The small residual strength then reduces more gradually until melt down at around 1500°C. For cold worked steels there is a more rapid decrease in strength after 300°C. The thermal properties of steel at elevated temperatures are found to be dependent on temperature rather more than stress level and rate of heating.

Table 5.11 Fire resistance of reinforced concrete (Courtesy IStructE)

	Nature of construction and materials		Minimum dimensions mm, excluding any finish, for a fire resistance of					
			½ h	1 h	1½ h	2 h	3 h	4 h
Slabs: ribbed open soffit	1	Reinforced concrete (simply supported)						
		(a) Normal-weight concrete — thickness	70	90	105	115	135	150
		width	75	90	110	125	150	175
		cover	15	25	35	45	55	65
		(b) Lightweight concrete — thickness	70	85	95	100	115	130
		width	60	75	85	100	125	150
		cover	15	25	30	35	45	55
	2	Reinforced concrete (continuous)						
		(a) Normal-weight concrete — thickness	70	90	105	115	135	150
		width	75	80	90	110	125	150
		cover	15	20	25	35	45	55
		(b) Lightweight concrete — thickness	70	85	95	100	115	130
		width	70	75	80	90	100	125
		cover	15	20	25	30	35	45
Walls	1	Less than 0.4% steel Normal-weight aggregate — thickness	150	150	175	200	—	—
	2	1% steel Normal-weight aggregate (concrete density 2400 kg/m³) — thickness	100	120	140	160	200	240
		cover	25	25	25	25	25	25
	3	More than 1% steel Normal-weight aggregate (concrete density 2400 kg/m³) — thickness	75	75	100	100	150	180
		cover	15	15	20	20	25	25
	4	Lightweight aggregate (concrete density 1200 kg/m³) (Note: intermediate densities may be interpolated.) — thickness	100	100	115	130	160	190
		cover	10	20	20	25	25	25

The Cardington fire tests proved that a steel-framed building designed for a specific resistance did not necessarily immediately collapse after the expiry of that time. This is due to structural continuity and inherent robustness of the frame. It is also apparent that a heavy, massive steel section will heat up more slowly than a light slender section. Modern fire engineering methods permit the calculation of the fire resistance of uncased steel.

Repairs to a fire-damaged steel-framed building cannot be contemplated until a forensic analysis of the fire has been made. This should include a history of the

Figure 5.27 Improving fire resistance of steel column by use of concrete brickwork.

fire and an assessment, if possible, of the temperatures reached and a survey of the residual distortion of the frame and its structural connections. Guidance can be gained from a study of well reported case studies.

Fire resistance to steel members, where required, can be in the form of board, spray or intumescent coating. It is claimed by the steel industry that fire protection is cheaper, in real terms, than it was 10–15 years ago. Intumescent coatings may now be applied (for new construction) in the fabrication shop. This material once occupied a niche market but is now universally accepted as a competitive method of providing fire protection. The fire resistance of new and existing steel can be enhanced by one of the following methods:

- Casing with fire–resistant boarding.
- Coating with fire-resistant sprays.
- Coating with intumescent paint. (Thin film intumescent coatings provide a good standard of finish and can be applied off-site. The ranges of fire resistance available are 30, 60, 90 and 120 minutes. For practitioners investigating existing installations it may be possible to locate a copy of a *Checklist of intumescents available in the UK* produced by British Steel but now discontinued. More recent information is available on SN19 10/2007.)
- Filling the spaces between flanges and webs with masonry – typically used for dealing with I-section stanchions (see Fig. 5.27).

A somewhat innovative method of dealing with fire in steel structures is to use a frame constructed using hollow sections and arrange for these to be filled with water or other similar fluid. This technique received a patent in America in 1884, although the first known use was in Pittsburgh in 1970. A building utilising this concept (Bush House, Cannon Street, London) was designed by ARUP and constructed in 1977.

5.3.6.5 *Fire in masonry structures*

Fire resistance of brick masonry structures is usually assessed on structural elements, such as block or brick walls and columns. There are no statuary requirements for individual units or the mortar used to construct walls or columns. In practice, masonry walls have demonstrated excellent fire resistance provided that the supporting structure maintains integrity for the duration of the fire.

In the absence of guidance in UK Codes useful information on this topic can be obtained from excellent publications by Edgell (1982) and de Vekey (2002). De Vekey gives guidance on the fire resistance of masonry walls for periods from 30 minutes to 6 hours. The fire resistance of masonry walls can be increased by applying insulating plasters or renders.

5.3.6.6 Fire in timber structures

Although timber is classified as a combustible material, a well-designed timber structure can perform well in a fire. Lightweight timber structures, for example in modern timber frame housing, will normally be fire protected by fire resistant cladding (e.g. plasterboard). Heavy timber construction has good inherent fire resistance due to the charring effect. When heavy timber members are exposed to a fire, the temperature of the fire-exposed surface of the members is close to the fire temperature. When the outer layer of the wood reaches about 300° C the wood ignites and burns rapidly. The burned wood becomes a layer of char which loses all its strength but retains a role as an insulating layer preventing excessive temperature rise in the core. The fire performance of timber is dependent on the charring rate and the loss in strength and modulus of elasticity. Examples exist of heavy timbers coated with intumescent paint to achieve the required fire resistance.

5.3.6.7 Effect of fire on glass

When glass is exposed to fire it is prone to failure due to imposed thermal stresses because of the rapid build-up of heat. Due to its brittle nature, failure will often be sudden. Borosilicate glass expands less than other glasses in the presence of heat and is therefore a more suitable material for use in circumstances where fire resistance is important. In older buildings, wired glass has been used to give increased resistance to fire but these materials have been largely superseded by laminated glass incorporating intumescent material. These materials offer better performance than wired glass. Glass, by itself, is not fire resistant. However it is non-combustible and there are actions in combination with other factors that can improve its performance in fire. Of importance is the type of glass, the framing, the beading, and frame restraint detail. The following gives some broad guidance on the performance of four types of glass:

- Wired glass: On exposure to fire, the glass breaks due to thermal shock but the wire mesh within the glass maintains the integrity of the panel by holding the fragmented pieces in place. Such glass is known as non-insulating glass.
- Special composition glass: On exposure to fire, the glass does not break owing to its low coefficient of thermal expansion and hence remains within its frame. The glass may also be thermally strengthened to minimise the

effects of stress, thereby achieving a level of safety from impact. Such glass is also known as non-insulating glass.

- Partially insulating glass: Has fire resisting properties that lie between insulating and non-insulating material. They are usually multi-laminated panes incorporating one intumescent interlayer which becomes opaque on heating. These can be utilised to achieve up to 15 minutes fire resistance.
- Insulating glass: Able to resist the passage of smoke, flames and hot gases and will meet the insulation criteria of at least 30 minutes. Two types of insulating glass are available. The first is intumescent laminated glass formed from multi-laminated layers of float glass and clear intumescent interlayers. The second is gel-interlayered glass formed from a clear, transparent gel located between sheets of toughened glass separated by metal spacers and sealed at the edges. The level of fire resistance achieved is related to the thickness of the gel interlayer.

5.3.6.8 Fire in plastics

Plastics perform poorly in fire as they are combustible. They emit noxious fumes, flaming droplets and melt quickly. A useful table indicating the performance

Table 5.12 Behaviour of common building plastics in fire (based on Lyons 2007, Courtesy Elsevier).

Material	Behaviour in fire
Thermoplastics	
Polythene/Polypropylene	Melts and burns readily.
Polyvinyl chloride	Melts, does not burn easily, but emits smoke and hydrogen chloride.
PTFE/ETFE	Does not burn, but at high temperatures evolves toxic fumes.
Polymethyl methacrylate	Melts and burns rapidly, producing droplets of flaming material.
Polystyrene	Melts and burns readily, producing soot, black smoke and droplets of flaming material.
ABS copolymer	Burns readily.
Polyurethane	The foam burns readily producing highly toxic fumes including cyanides and isocyanates.
Thermosetting plastics	
Phenol formaldehyde Melamine formaldehyde Urea formaldehyde	Resistant to ignition, but produce caustic fumes including ammonia.
Glass-reinforced polyester (PRP)	Burns producing smoke, but flame-retarded grades are available.
Elastomers	
Rubber	Burns readily producing black smoke and sulphur dioxide.
Neoprene	Better fire resistance than natural rubber.

of several types of plastics can be found in *Materials for Architects and Builders* (Lyons 2006). Table 5.12 shows the behaviour of some common building plastics in fire.

5.3.7 Vibration

In refurbishment schemes it may be essential to check that the re-arrangement of accommodation is not detrimental to the vibration characteristics of the structure. Partitions may add considerably to the damping of vibration so new arrangements need to be checked for their effect on the system. Particularly sensitive are hospital operating theatres. Retrofitting to improve vibration characteristics is possible but requires expert assessment to obtain good results. Testing can be carried out to check performance of existing or new installations but is expensive. At the time of writing, the typical cost of testing was in the region £10–15k.

Factors influencing human perception of vibration and associated noise include:

- type of activity
- time of day when the activity is being undertaken
- type of environment where activity is taking place
- direction of the vibration
- amplitude of vibration (usually less than 1 mm)
- frequency of vibration
- source
- duration of the exposure

In general, machinery should be mounted using resilient support systems.

5.3.8 Workmanship and site practice

Good workmanship is an essential ingredient of good construction. Such a standard can only be achieved if the details are easily constructed, the materials are of good quality and the quality, training and experience of the work force is appropriate. For the refurbishment and repair of historic structures above average skills may be required. For example the restoration of medieval timber frames may require carpenters with a knowledge of ancient jointing of timber.

The general skill of the British workforce has probably diminished in recent times due to the scarcity of first class apprenticeships. This scarcity has only recently been addressed although artisans from Eastern Europe are partially filling the gap. Common shortcomings in workmanship and site practice are detailed below.

5.3.8.1 General

- Movement joints bridged over.
- Sealing of movement joints with materials that transmit too much load.
- Structures built out of plumb thus invalidating cladding fixings (all multi-

storey frames must be surveyed before cladding commences, thus allowing, where necessary, dimensional adjustments to fixing details to accommodate frame inaccuracies).

5.3.8.2 Timber

- Poor site storage of trussed rafters and other timber materials.
- Use of poorly seasoned timber.

5.3.8.3 Concrete

- Misplaced reinforcement in RC structures due to bad concrete placing techniques.
- Poor detailing and scheduling of reinforcement.
- Poor control of cover to reinforcement.
- Misplaced reinforcement due to bad fixing practice.
- Insufficient curing of green concrete particularly in extreme temperatures.
- Poor compaction of concrete.
- Over vibration of wet concrete.
- Badly constructed formwork.

5.3.8.4 Masonry

- Bricks in load-bearing masonry laid frog down.
- Weep holes in masonry omitted or badly formed.
- Incorrect mortar mix.
- Poor tie-spacing.
- Inadequate or misplaced ties.
- Mortar droppings in cavities.
- Incorrect bonding.
- Insufficient bedding.
- Incorrect cavity width and tie embedment.

5.3.8.5 Steel

- Bolts missing from steel frames.
- Poor quality welds.

5.3.8.6 Glass and glazing

- Poor installation of glass, particularly double glazing. (Attention is drawn to the FENSA self assessment scheme set up by the Glass and Glazing Federation which deals with good practice in the installation of replacement doors and windows.)
- Inadequate thickness of glass.
- Wrong specification of glass in high fire risk situations.

5.3.9 General repairs

BRE promote a series of publications under the generic title *Good repair guides*. These include:

GR1.1996	*Cracks caused by foundation movement*
GR2.1996	*Damage to buildings caused by trees*
GR3.1996	*Repairing damage to brick and block walls*
GR4.1996	*Replacing masonry ties*
GR5.1997	*Diagnosing the causes of dampness*
GR6.1997	*Treating rising damp in houses*
GR7.1997	*Treating condensation in houses*
GR8.1997	*Treating rain penetration in houses*
GR9.1997	*Repairing and replacing rainwater goods*
GR10.1997	*Repairing Timber Windows*
Pt1.	*Investigating defects and dealing with water leakage*
Pt2.	*Draughty windows, condensation in sealed units, operating problems, deterioration of frames*
GR11. 1997	*Repairing flood damage*
Pt1.	*Immediate action*
Pt2.	*Ground floors and basements*
Pt3.	*Foundations and walls*
Pt4.	*Services, secondary elements, finishes, fittings*
GR12. 1997	*Wood rot: assessing and treating decay*
GR13. 1998	*Wood-boring insect attack*
Pt1.	*Identifying and assessing damage*
Pt2.	*Treating damage*
GR14. 1998	*Recovering pitched roofs*
GR15. 1998	*Repairing chimneys and parapets*
GR16. 1998	*Flat roofs*
Pt1.	*Assessing bitumen felt and mastic asphalt roofs*
Pt2.	*Making repairs to bitumen felt and mastic asphalt roofs*

5.3.10 Stability and robustness

When refurbishing or remodelling a structure it is important that the revised structure is at least as robust as the original. In general it is apparent that framed buildings are more robust than those without a frame. An interesting comparison was made when gas explosions of similar intensity occurred in two similar high rise blocks. In one, a structure of loosely tied precast reinforced concrete units (Ronan Point), the outcome was a disproportionate collapse and subsequent demolition. The block with an *in-situ* frame (Mersey House) suffered damage that was largely confined to one apartment, readily repaired and re-commissioned.

Technical papers have drawn attention to the bookend effect (progressive longitudinal distortion of certain 19th century, Georgian and Victorian straight terraces caused by cyclical expansion and contraction of their continuous facades). The possibility has also been raised of domino-type collapses occurring if too much of the buttressing structure is removed when remodelling cross-wall construction. Cases have been reported of shops with housing above and other housing collapsing in Islington, London for lack of attention to these considerations.

Structural engineers will be familiar with the term 'redundancy'. This usually indicates the provision of additional members in a structure to those necessary to carry normal service loads. Should the structure then be subjected to exceptional or abnormal loads, the additional members can assist in carrying those loads safely to the foundations albeit with some distortion in the structure. In that way disproportionate collapse will have been avoided.

Although stability is important for permanent structures, the need for stability in temporary structures or those under erection, is equally important. BCSA have issued guidelines for the *Stability of Temporary Bracing* in which they outline the following factors to underpin a safe system of working:

- A sound plan – a written erection method statement forming the basis of a safe system of work.
- Adequate resources – people, suitable equipment selected and tested.
- Competent individuals – selection of suitably trained and experienced operatives and supervisors.
- A chain of command – enabling clear instructions and briefings to be given.

They propose three objectives to be addressed in the method statement:

- To ensure individual pieces and the part-erected structure stands up throughout the construction stage.
- To operate cranes and other plant to lift and position safely.
- To provide safe working positions for erectors and safe access to/egress from those positions.

5.3.11 Façade retention

Façades of buildings may need to be retained for a variety of reasons including:

- English Heritage requirements
- Client's instructions
- Aesthetic reasons

Such façades may need support during the building operation, in which case an appraising engineer should carefully check the condition of the façade and arrange to repair any fabric damage and brace window and door openings.

The façade should then be temporarily stabilised with a braced structure. This is usually constructed using structural steel sections or scaffold fittings (see Figs. 5.28–5.32). Occasionally timber may be suitable. Where loads are supported at ground level an adequate site investigation must be carried out to check the competency of the soil or paving to carry the loads.

The restraining structure must not be removed until sufficient permanent support and restraint is provided by the new structure. Designers should make adequate allowance for possible differential settlement between old and new work. Although unlikely, it is possible (where internal planning permits) to introduce sufficient new structure behind the façade to render a temporary support unnecessary. Useful guidance for the design of structure to retain façades is given in the bibliography.

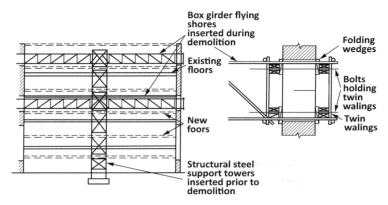

Figure 5.28 Façade retention (1) (Redrawn based on CORUS).

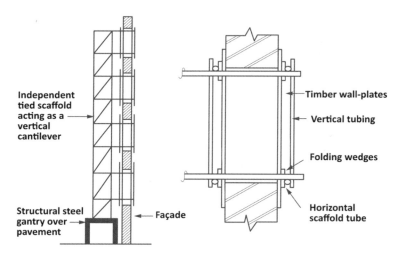

Figure 5.29 Façade retention (2) (Redrawn based on CORUS).

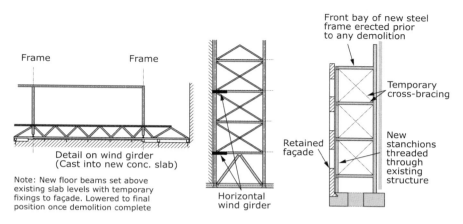

Figure 5.30 Façade retention (3) (Redrawn based on CORUS).

Façade tied to column

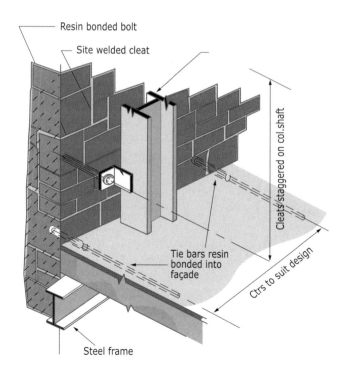

Figure 5.31 Façade retention (4) (Redrawn based on CORUS).

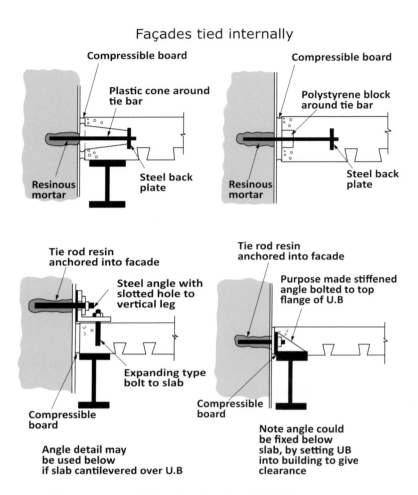

Figure 5.32 Façade retention (5) (Redrawn based on CORUS).

5.3.12 Foundations

5.3.12.1 General

(Including contaminated ground and brown-field sites: see also website for contaminated land www.geocontaminatedland.co.uk).

The UK government defines a brown-field site as land that is or was occupied by a permanent structure (excluding agricultural or forestry buildings) with associated fixed surface infrastructure. The definition covers the curtilage of the development and includes land used for mineral extraction and waste disposal where provision for restoration has not been made through development control procedures. (This would appear to exclude open cast coal sites where it is customary for the extraction contractor to reinstate the site before completion of contract.)

Typical hazards encountered in redeveloping such sites include:

- ground movement
- contaminants
- durability and serviceability of construction materials
- gas migration
- subterranean fires
- old mine workings

Such hazards can only be defined with the benefit of well-designed and executed site investigations. The investigations will help to advise the developer of the appropriate remedial measures to restore the site to a low risk usable state. Practitioners must guard against undue pressure from developers to recklessly minimise remedial work in order to reduce costs. Remediation techniques include:

- preloading (see Section 5.3.12.3)
- dynamic compaction (see Section 5.3.12.3)
- vibrated stone columns (seeSection 5.3.12.3)
- cover layers and containment barriers (the provision of horizontal blankets and vertical cut off walls to prevent migration of contaminants)
- soil mixing
- stabilisation
- grouting
- excavation and disposal
- excavation and re-compaction
- excavation and physical treatment of the spoil
- leaching, washing and flushing
- soil vapour extraction and air parging (the removal of contaminants that can be carried by air)
- groundwater treatment
- thermal processes
- chemical treatment
- electrical treatment
- phytoremediation (the use of plants to treat soil and groundwater contaminants)
- natural attenuation

(These processes are fully explained in BRE Report BR485 *Brownfield sites*.)
There are several types of foundation including:

- strip (including trench fill)
- pad
- raft (including piled raft)
- piled (piles may be bored or driven)

- those that rely on consolidated ground including backfilled former mine workings

Foundations may have been constructed on excessive depths of fill material leading to unacceptable differential settlement and damage to underground services. The general advice given for the construction of ground floor slabs, particularly for domestic housing, is to limit compacted made-up ground to a maximum depth of 2 m: beyond that to use suspended construction.

In some situations chemicals (e.g. sulfates) have been sucked from the ground by an action known as wicking.

In situations where new (and deeper) foundations are to be constructed adjacent to existing foundations it may be necessary to underpin the existing foundations to avoid a surcharge from the new work. Underpinning may be achieved by deepening existing strip footings using the hit and miss process or, nowadays, by needle piles inserted diagonally beneath the existing footings. For detailed advice consult specialist literature.

Foundations constructed on sensitive clays may be affected by seasonal settlement and/or heave. These clays lie predominantly (but not entirely) below a line stretching from the Wash to the Bristol Channel. Typically these may be London, Gault, Weald, Kimmeridge Oxford or Lias clays together with clays from the Woolwich and Reading beds. They may show marked swelling with increase of moisture content, followed by shrinkage after drying out. In general the leaner glacial clays are not so affected. However these leaner clays can show substantial shrinkage if influenced by the roots of growing trees. There is no simple field or laboratory test to identify shrinkable clays although as a rule of thumb clays with a liquid limit in excess of 50 should be regarded with suspicion.

Since about 1950 foundations for non-sensitive structures have been placed at 0.9 m into natural ground and performance has been generally satisfactory. Care should also be taken to use suspended ground floor construction where potentially shrinkable clays are found. In such cases it is customary to leave a void beneath the slab to allow some seasonal movement of the soil to take place. Proprietary systems are also available to accommodate movement of the soil.

In cases where damaging shrinkage or heave has occurred it may be necessary to underpin the existing foundations (see Section 5.3.12.2).

If the likelihood of sulfate damage is suspected the use of sulfate-resisting cement (SRC) is recommended for concrete foundations.

5.3.12.2 Underpinning

Underpinning is a construction used to deepen, widen or restore foundations. The need to do this may arise from defective foundations, for example settlement in sensitive clays, or in a situation where existing foundations might damagingly surcharge the foundations of a nearby new structure. There is a number of methods of underpinning which include:

Figure 5.33 Single-storey building underpinned using Abbey Pynford composite units (Courtesy Abbey Pynford).

Traditional

This describes the construction of a new continuous foundation (constructed by hit and miss methodology). This can usually be carried out from one side of a wall.

Beam and pier

In this method a beam in a load-bearing masonry wall is inserted to support the wall. The ends of the beams are supported on piers taken down to competent load-bearing ground. Access to both sides of a wall may be required for this type of underpinning although new techniques may remove the need for this.

Beam and pile

This method is similar to beam and pier but with the beams supported by new piles. These piles may be conventional or mini-pile.

Mini-piles

It is stated that Fondedile first patented the technique of drilling through a masonry foundation into the ground beneath, inserting reinforcement and then grouting up.

Cantilever slab

In this method a thickened ground slab is supported by piles usually placed close to the inner face of the internal walls and cantilevers project into the walls at approximately 1 m centres to support them.

Enlarging foundations without deepening

This is a technique for extending existing foundations by lateral additions to the foundations and ensuring that the old and new foundations work together in harmony to share the imposed loads.

5.3.12.3 Ground improvement

The bearing capacity of soils may be enhanced by a variety of methods which include:

Dynamic compaction

This is essentially achieved by dropping large weights on to the soil to be treated. This is a somewhat crude procedure and must be verified by subsequent soil testing to check the validity of the treatment.

Vibro-compaction

This is a technique that establishes a series of strategically placed stone columns to compact the ground. It is most effective in non-cohesive soils.

Preloading

In this technique the area to be treated is surcharged with a layer of soil which is left in place for a specified period and then removed.

5.3.12.4 Mining subsidence

Subsidence may be caused by the below ground extraction of minerals and other materials. These include:

- Coal (once very extensively worked but now considerably depleted with the demise of much of the coal mining industry).
- Brine (pumping from salt-bearing rock in Cheshire has caused long term subsidence over a wide area).
- Oil or gas (although most UK oil or gas deposits have been under the North Sea or in Morecambe Bay, some deposits have been found inland).

The extents of these materials, where known, have been extensively plotted and can be observed on maps available from specialist surveyors.

Practitioners may need to have a working knowledge of these activities when considering extensions and/or refurbishment of building structures.

Two main methods of extraction were used:

Pillar and stall

In this system pillars of coal were left standing and the coal extracted around them. The voids thus formed were either left empty or backfilled with suitable material. Sometimes this material was the waste arising from coal-washing. Sometimes large pillars of coal were left under churches and other important buildings.

Longwall working

In this system coal would be extracted over a wide advancing front. Temporary props would be removed after a period of time allowing a subsidence wave to pass over the site. This method had the advantage that once subsidence had taken place no cavities would exist.

When coal extraction was nationalised, considerable data was available from the NCB at their London headquarters. Their handbook published in 1974 (still available from Institution libraries) contains much relevant advice including the following general commonsense principles:

- Structures should be completely rigid or completely flexible.
- The shallow raft foundation is the best method of protection against tension or compression strains in the ground surface (as a result many houses or similar light structures have been successfully built using a 150 mm thick reinforced concrete raft (see Tomlinson 2001, p.207 for detail of the raft) as recommended by the then Ministry of Works.
- Large structures should be divided into independent units. The width of the gaps between the units to be calculated from a knowledge of the ground strain derived from the predicted ground subsidence.
- That small buildings be kept separate from one another, avoiding linkage by connecting wing walls, outbuildings, or concrete drives.

It is worth noting that some high rise buildings, possibly subject to future mining settlement, have been constructed on split cellular reinforced concrete rafts with jacking pockets to permit re-levelling in the event of differential movement.

In some areas extensive backfilling with fly ash and other materials will minimise the possibility of settlement when building over former workings.

5.3.13 Defective basements

Basements may be defective for a variety of reasons which include:

- structural problems
- dampness or penetration of water vapour
- breakdown of tanking
- defective design and/or construction of reinforced concrete construction
- leakage in the clutches of steel sheet piling due to non-existent or poor caulking
- inappropriate choice of construction, for example masonry used in an area of high or rising ground water.

Before proceeding with repair or refurbishment it is vital to understand the potential use of the basement. This can conveniently be assessed against a grading indicated in BS 8102 and other authoritative documents on the topic. These grades are as follows:

(1) Basic utility, e.g. car parking; plant rooms (excluding electrical equipment), workshops. (Some seepage and damp patches are tolerable.)
(2) Better utility, e.g. workshops and plant requiring drier environment, retail storage. (No water penetration but moisture vapour tolerable.)
(3) Habitable, e.g. Ventilated residential and working, including offices, restaurants, leisure centres. (Dry environment essential.)
(4) Special archives and stores requiring controlled environment. (Totally dry environment essential.)

Each grade requires a particular construction varying, for example from un-tanked reinforced concrete for Grade 1 to reinforced concrete together with drained cavity arrangements for Grade 4. Similar designs using masonry or steel sheet piling may also be appropriate to match the requirements of the various grades.

Tanking should normally be applied externally and may take the form of asphalt (rare these days) or polymer sheeting with a bituminous coating. Should the tanking be found to be damaged it may be possible to excavate down to the affected area in order that repairs may take place.

In planning repairs it should be noted that ground water levels are rising and it may be that provisions for resisting the entry of water should be made more stringent than required in the original construction. Also it is necessary to check against possible floating in the new situation. It is prudent to design all below-ground basement walls for water pressure of at least a third of the height of the wall even though no water (or risk of it) is indicated in the site investigation.

Probably the most effective method of repair (if space permits) is to construct a system of drained cavities and drain away any penetrating water using automatically operated pumps working in sumps.

5.3.14 Liquid retaining structures

The construction materials most commonly used to contain liquids are steel and concrete (reinforced and/or prestressed). A distinction must be made between liquid retaining and liquid excluding structures. Liquid excluding structures are dealt with in Section 5.3.10. Unless the design and construction is to an appropriate standard, leakage may occur.

Steel structures (tanks, vats etc) are relatively easy to repair. The location of the leak is usually obvious and repairs can be effected by welding additional plates across the defective area.

Concrete tanks must be designed to exacting standards (see BS 8007:1987 *CP for the design of concrete structures for the retaining of aqueous liquids*). If properly designed and constructed reinforced and/or prestressed concrete can be a very effective and economical medium for retaining liquids. However faults do occur and may be due to one or several of the following:

- Poorly constructed movement or construction joints.
- Inadequate provision of movement joints.
- Porous concrete due to inappropriate mix and/or poor compaction.
- Inadequate curing allowing cracks to develop.
- Misplaced water stops.
- Corrosion of reinforcement due to poor steel fixing allowing inadequate cover or carbonation within the expected life of the structure.
- Breakdown of special protective linings or coatings.

Repair solutions to be considered include:

- Sealing cracks with high pressure epoxy based material.
- Lining the inside of the tank with a water resisting material such as a butyl based polymer.
- Demolishing and re-building the tank to a higher standard.

5.3.15 Explosions in structures

Two of the most far-reaching incidents of explosions in buildings were:

- Ronan Point 1968 (a block of flats).
- Abbeystead 1984 (a water transfer scheme between the rivers Lune and Wye in Lancashire).

Both were the result of gas leaks. Ronan Point was caused by a relatively small intensity town gas explosion on the 18th floor of a 22-storey block of flats. The flats were constructed a few months earlier using the Larsen Neilsen large precast concrete panel system and resulted in the disproportionate collapse of all 22 stories of a corner of the block. An explosion of similar intensity in the ground floor of a 16-storey block in Bootle, Lancashire, built using an *in-situ* reinforced concrete frame sustained damage but this was limited to the area immediately adjacent to the explosion. As a result of Ronan Point the building regulations were modified in an attempt to limit damage in a re-occurrence of such an incident.

Abbeystead was the result of an ignition of a mixture of methane and air which had accumulated in the wet room of the Valve House. Sixteen people were killed and others injured. Although perhaps unique to the UK, this type of incident was not unknown to engineers working on hydro-electric schemes where concentrations of methane and air were sometimes to be found in river diversion tunnels. The HSE enquiry report into Abbeystead contained a number of design recommendations of which perhaps the most important was; Systems conveying water should be so designed that any gas discharged, either during filling or at any other time, is vented to a safe place in the open air.

Practitioners involved in the design and/or construction or renovation and repair of structures similar to the above should carefully heed the advice promulgated as a result of the above incidents. Particular care should be taken in designing or modifying basements.

It should be noted that explosions due to terrorist actions are not catered for in the building regulations.

5.3.16 Radon gas

Radon is a colourless, odourless naturally occurring, radioactive gas which poses a health risk to a relatively small number of people in their homes. Radioactive decay of uranium generates radium which in turn decays to radon. Very small but variable quantities are found in all soils and rocks but it is commonest in areas underlain by granite or limestone. Radon levels vary throughout the country and can differ between adjacent buildings. The National Radiological Protection Board (NRPB) has recommended levels of concentration above which preventative action should be taken. Recent reports from the Health Protection Agency (HPA) suggest that 100,000 homes are at risk from radon.

Simple and relatively inexpensive tests are available to check whether existing properties are above the level at which remedial action should be taken. Such action may include:

- Installation of a radon sump system.
- Improvement of ventilation under timber suspended floors, (e.g. the installation of air bricks in the substructure walls).
- Installation of positive ventilation in the property.
- Sealing of all cracks and gaps in solid concrete floors (suspended and ground bearing floors).
- Changing the way properties are ventilated (may be an expensive option).

For new properties (and even those built in the last 15 years or so) there is a requirement to comply with building regulations. The level of precautions recommended (basic to full) varies in relation to the perceived risk. BRE provide a Radon Hotline and enquiries may be made to partscdgh.br@odpm.gsi.gov.uk and through the Department for Environment, Food and Rural Affairs (Defra).

5.3.17 Impact damage

Buildings and structures may suffer impact damage from rail, road and marine traffic also construction operations: injury or loss of life may result. The repair or replacement of structure will result in commercial loss.

As a consequence the construction of additional structure to prevent accidental damage is now regularly considered.

One of the most spectacular accidents was the 1957 Lewisham Disaster when two trains collided bringing down an over-bridge. Footbridges over main roads have been demolished by excessively high vehicles passing beneath. The construction industry has been slow to react but since 1987 BS5400 has required the consideration of collision loads when designing a bridge.

Another area of concern is the impact of road vehicles on parapets. Again London Underground (LUL) has been the victim of examples of this phenomenon. In 1978 a fire engine demolished the parapet of a bridge carrying the Rotherhithe New Road over the East London Line. More dramatically, in September 2008 the M8 motorway was plunged into chaos after a flatbed trailer carrying a refuse lorry struck a bridge. The lorry was torn from the back of the vehicle.

Guidance on design forces to cater for these situations may be obtained from DoT/HA standards. Analytical techniques for modelling the behaviour of masonry walls in such circumstances have been developed at the University of Liverpool.

5.3.18 Flat roofs

In a comprehensive research document for the Chartered Institute of Building (CIOB) Brian Barnes examined 304 reports of flat roof defects. His analysis indicated that on average 70% of problems stemmed from lack of:

- knowledge
- care
- skill
- supervision

Barnes quoted a DoE/PSA report which identified the main faults as follows:

- deterioration of the membrane
- distortion of box guttering
- cracking of membrane
- corrosion of lead or zinc sheeting
- water ingress at fixing points
- water ingress at lap joints
- condensation on underside of profiled metal sheeting
- corrosion of steel sheet roofing
- insulation core of sandwich panels attacked by birds
- failure of movement joint
- failure of timber roof deck
- cracks in mastic asphalt/ lead junction
- water ponding on flat roofs (except where a roof was deliberately designed for ponding to protect surface materials). (This is usually caused by excessive deflection of the deck or if the original falls were laid to around 1:80. The current recommendation is for the falls to be 1:40 to allow rapid disposal of rainwater.)

- water backing-up down-pipes
- degradation or collapse of insulation
- water penetration from brick upstands/parapets/copings
- condensation on rainwater pipes in roof void
- spalling of concrete at edge detail
- failure at abutment lead flashings
- blistering, cracking, slumping and cockling in asphalt
- surface degradation of asphalt
- delamination of protective aluminium foil from bitumen backing sheet
- wind scour of gravel ballast
- loss of adhesion of felt membrane
- splits in roof membranes at edge trim
- degradation of felt membrane under joints in paving slabs

Obviously this list (which is not comprehensive) deals with several different types of construction. However it reinforces the view that a very detailed structural and condition survey report is an essential precursor to commencing any remedial work. One of the most repetitive faults in flat roofing is making insufficient allowance for storm-water drainage. This often occurs where no allowance has been made for the structural deflection of the roof under working load.

It should be noted that the use of cold deck construction is no longer allowed in the UK due to the risk of interstitial condensation (see Section 5.3.18.1).

Repair work will be related to the actual defects discovered; in an extreme case it may be advisable to superimpose a pitched roof over the original construction. This technique has been adopted on a number of domestic dwellings built in the post-Second World War boom.

5.3.18.1 Cold deck construction

This is a traditional form of flat roof construction. It is, however, no longer a recommended method of construction of flat roofs in the UK. By the early 1970s, for example, it was already banned in Scotland.

Cold-deck flat roof construction means that the insulation is placed below the deck. This is in contrast to conventional warm-deck or inverted warm-deck flat roofs where the insulation is placed above the deck.

In concrete slab cold-deck flat roofs the insulation therefore is placed on the underside of the slab – i.e., its soffit. On the other hand, in timber deck cold roofs the insulation is placed between the joists, with a recommended minimum 50 mm air space needed above the insulation for ventilation. This air space, however, can easily become compromised by debris or waste such as a wasps' 'bike' or 'byke' (the name 'bike' refers to a paper-like substance made mainly from wood fibres gathered locally from trees and chewed by the wasps and softened by their saliva).

In both cold-deck forms, the roof construction above the insulation is inevitably not kept reasonably warm especially during the winter period. Any warm moist air from the room below that may permeate through the insulation will condense on relatively cold surfaces into which it comes into contact. In other words, there is a risk of interstitial condensation occurring in the roof structure when the temperature of the construction above the insulation drops below the dew point.

As part of refurbishment work to a building it is often appropriate to upgrade the thermal and weathering performance of its flat roof. Before doing so, therefore, it is essential to establish whether or not it is of cold-deck construction. This can be done by uplifting a small (say, 0.5 m²) section of the deck if it is timber (or exposing the slab soffit if it is concrete) to check for the tell-tale presence of insulation.

If it is a cold-deck roof, then the original insulation should be removed before installing the new roof covering and insulation using the warm deck construction system. This is to avoid creating a varied and therefore complex environment in the roof, which could result in condensation occurring in the unventilated space between the old and new constructions. It is not necessary, however, to ventilate a warm deck flat roof. The only exception to this is in the case of insulated lead-sheeted roofs, which require a 50 mm vent space above the insulation to minimise the risk of underside lead corrosion (Lead Sheet Association, 1990).

A major problem during the renewal/replacement of a roof covering, whether flat or pitched, is ensuring the building's continued weather-tightness. This is even more difficult to achieve when converting a timber cold deck flat roof into a warm deck one. In such a case the space between the joists and the ceiling finish below would be exposed, albeit temporarily, to the elements. A sudden, unexpected downpour of rain could cause extensive water-damage to the ceiling and room below. To minimise if not prevent such damage some form of interim roof frame protection covered with tarpaulins or other heavy duty temporary weatherproof sheeting securely anchored should be installed over the exposed parts of the roof at the end of each working day.

Interstitial condensation, however, is not the only defect affecting flat roofs. Tables 5.13(a)–(d) summarise the problems that can be found in flat roofs generally.

Table 5.13(a) Defect Diagnosis Checklist – interior (based on Euroroof Ltd. 1985).

Defect	Likely cause	Remedial action
Leaking and staining – External wall and junction with ceiling.	Failure at parapet wall skirting, gutter, eaves or verge finish on roof. Possible leak from services.	Record findings and inspect externally for actual cause. Consider replacing insulation and membrane with new warm deck roof system (see Douglas, 2006). Alternatively, repair leak if this is the cause.
Leaking and staining around rooflights and openings.	Failure at junction of rooflight or upstand flashing. Broken glass, defective seals between glass and glazing bars or frame.	Replace defective rooflight or its glazing in its entirety.
Leaking and staining in the main body of the ceiling.	Rainwater penetration through a damaged membrane or serious interstitial condensation.	Replace insulation and membrane with new warm deck roof system.
Leaking and staining at junction with structural supports.	Movement at construction joint.	Make good or improve expansion joints.

Table 5.13(b) Defect Diagnosis Checklist – roof edge (based on Euroroof Ltd. 1985).

Defect	Likely cause	Remedial action
Parapets – Dislodged or distorted coping, with no dpc below, brick spalling.	Thermal or moisture movement, wind or mechanical damage, frost or sulfate attack.	Rebuild affected parts of wall, repoint open or unsound joints in copings or walling.
Parapets – Blistering, sagging, torn, cracked, crazed or split skirting, open lap joints.	Thermal or moisture movement of backing, lack of adequate key to backing.	Remove and replace by specialist contractor providing good key in base.
Skirting less than 150 mm in height.	Design, workmanship or supervision fault.	Renew if practical to correct dimensions.
Defective expansion joints.	Too wide spacing between joints, poor materials or workmanship, ageing of sealant.	Renew sealant and modify or renew flashings at junction of parapet with roof expansion joints may be necessary.
Gutters and outlets – Blocked outlets.	Lack of inspection and poor maintenance. Outlets too close to skirting.	Clear blockage. Should be resealed by a specialist contractor.
Eaves and verges – Peeling or stripping of felt roofing and board insulants.	Wind damage due to poor edge detailing. Either inadequate bonding or mechanical fixing.	Strip and relay damaged roofing to improved edge design.
Eaves and verges – Cracks and splits in roof covering.	Differential thermal or moisture movement or settlement and lack of provision for these movements.	Strip affected roofing and insert minor movement joint of high performance felt over joint in base, bonded only at the edges.

Table 5.13(b) Defect Diagnosis Checklist – roof edge (based on Euroroof Ltd 1985) *(continued)*.

Defect	Likely cause	Remedial action
Eaves and verges – Cracks relating to metal edge trims	Differential movement between metal and waterproofing.	Replace with non-metal trims – ie, pultruded GRP, reseal with high performance felt cappings.
Fixings – Cracks or splits in roof covering around guard rails, eyebolts and fixings for roof tracks, jibs and lightning conductors.	Inadequate fixing, differential thermal or moisture movement between roof deck and waterproof membrane, shrinkage of roof covering due to ageing or lack of adequate protective finish	Remove and reset more securely or change to a more suitable design such as waterproof plinths.

Table 5.13(c) Defect Diagnosis Checklist – elements above roof (based on Euroroof Ltd 1985).

Defect	Likely cause	Remedial action
Windows and doors – Weakness in watersealing functions at jambs or sills of door and window openings in walls above roof level.	Inadequate design or faulty installation.	If seriously defective remove affected doors and windows, correct the watersealing arrangements, make good and refix. If less serious it may be possible to overcome the defects by fixing cover beads bedded in mastic between the frames and wall to bridge gaps between frames and DPCs.
Rooflights and openings – Defective seals and joints between opening lights and base. Defective dome light.	Thermal or moisture movement, frost attack, absence or water bar or sill drip, insecure fixing of frame, mechanical or wind damage to flashing, poor workmanship.	Depending on fault, make good fixings and flashing, rake out open joint and seal with elastomeric sealant, or fit draughtstrip to frame or if rotten renew frame. Increase drip and fit gasket or lead or high performance felt cover flashing to kerb.
Vents and flues, plant and equipment – Discolouration, softening, cracks and deterioration.	Condensed solvents or chemical attack.	Repair roofing using specialist roofing contractor. Raise vent to permissible height, or re-site it.
Vents and flues, plant and equipment – Early localised roofing failure.	Aggressive solvent or chemical attack.	Use specific protection method and/or specialist resistant membrane.
Expansion joints – Unsealed joints and screw head fixings in metal cappings.	Poor design or workmanship, use of butt or lap instead of welted or sleeved joints, poor detailing at ends or abutments.	Either remove capping and sealant between the loose jointing sleeve and capping and then fix with proprietary sealed head screws or install self-adhesive bitumen membrane immediately beneath capping.

Table 5.13(d) Defect Diagnosis Checklist – main area of roof (based on Euroroof Ltd 1985).

Defect	Likely cause	Remedial action
Cracks, tears and splits along line of support of roof deck.	Thermal or moisture movement or sagging of roof deck, possible saturation of insulation. Opening of joints in *in-situ* slabs.	Cut back felt over crack, dry base and edges of existing felt. Apply layers of a high performance felt, allowing 150mm overlap, the first layer bonded to base only at the edges and the top layer continuously bonded over the first layer and edges of existing felt.
Cracks, splits, cockles, ripples and rucks not along line of support.	Differential thermal or moisture movement between substrate and membrane. In particular shrinkage cracks in screed untapped joints.	If minor, treat as above. If severe, re-roofing of whole roof may be necessary.
Surface crazing, pimpling and crocodiling.	Lack of solar protection. Possible chemical attack.	If the effect is limited to the surface no treatment is necessary. If deep cracks are found or clear signs of chemical attack, re-roofing and special protection are indicated.
Blisters.	Water vapour pressure below the roof covering. Sometimes aggravated by lack of solar protection.	Star cut blister, rebond to underlay or base, patch over affected area. Before patching any entrapped moisture must be released and the roof dried out.
Interlayer blisters.	Moisture or contaminant between roofing layers.	Treat as blisters.
Punctures and rips.	Impact damage by foot traffic, other trades, birds or hail stones.	Patch repair as necessary. If possible limit access or provide walkways and protective surface.
Serious ponding.	If not a blocked outlet could be collapsed or rotten insulant, deflected decking, building subsidence or bad original design.	Remove affected insulant or deck. Renew deck and insulant and repair roof. If extensive may have to re-roof, including insulation tapered to 1:40 fall if practical.

5.3.19 Pitched roofs

There are three main categories of defects that affect pitched roofs. These are listed in Table 5.14.

5.3.20 Windows and doors

Replacing or repairing windows and doors is a common feature of most domestic and many commercial refurbishment projects. The most common materials for use in replacements are timber, uPVC and aluminium. Table 5.15 summarises these.

Table 5.14 Typical pitched roof defects.

Defect	Likely cause	Remedial action
Roof coverings		
Leaks	Any of the causes listed under this section	Depends on the extent of leakage. If localised, a patch repair may be sufficient. 'Turnerisation'* is not recommended as a long term solution (Douglas 2006).
Loose and broken slates/tiles	• Wind damage • vandalism • nail sickness • lack of maintenance • damage by careless maintenance operatives	Patch repair, if not replace entire roof covering.
Delamination of slates/tiles	• Frost attack, particularly on old slates/tiles • Acid rain pollution	Replace roof covering completely using copper fixings and breathable felt underlay.
Organic growths (eg, algae, moss, etc) on slates/tiles	• Exposure to vegetation nearby	Remove using a biocide applied with a power-jet spray.
Projections		
Leaks	• Defective or inadequate seals around collar/flashing.	Replace seal or collars.
	• Too many projections such as vent pipes too close together	Insert all projections into one main outlet box.
Misalignment	• Badly installed • Vandalism • Storm damage	Refix/resecure faulty vent pipes etc.
Roof structure		
Roof spread	• Defective or missing ties/rafters	Strengthen roof structure with additional struts, ties and/or purlins.
Roof sag (across rafters or along ridge)	• Built-in distortion • Struts removed/damaged • Timber decay in rafters, struts or ties	Leave but monitor annually. Replace defective sections.

* Turnerization is the name of a proprietary process of applying a layer of Hessian/fibreglass in a coat of bituminous paint to the external surface of a slate or tiled roof to enhance its weather tightness. It was derived from a process used by a company called Turner (Hollis, 2005).

Table 5.15 Typical defects in windows and doors.

Defect	Likely cause	Remedial action
Broken glass or sheet panels	• Vandalism • storm damage • user abuse • accidental impact damage	Replace broken glazing or panelling.
Condensation in double glazing units	• Breach in seal surrounding the double glazed pane.	Replace glazing with new sealed pane.
Corrosion of metal framing or sashes	• Inadequate rust protection and poor paint coating • Exposure to excessive levels of moisture	Remove loose rust and scale. Treat bare surfaces with a rust-inhibiting primer. Apply two-coat metallic/polymeric paint system.
Crazing/cracking of uPVC framing or sashes	• Oxidation accelerated by ultra-violet radiation.	Replace affected window or door units.
Decay of timber framing or sashes	• Poor or inadequate levels of paint protection. • Lack of drips, bevels and throatings in sashes and frames. • Exposure to excessive levels of moisture.	Patch repair using an acrylic filler if decay is localised. Otherwise replace whole unit with treated and painted frame and sash/door.
Malfunction of sash/ door	• Too much paint • Poor fit • Structural movement around opening	Remove excess paint. Adjust/replace defective ironmongery. Replace defective sash – but there might be problems in matching metric to imperial sizes.
Safety hazard	• Sill to windows on all upper floors less than 1.05 m above floor level. • Safety catch or other ironmongery not operating properly.	Install safety bar across window opening at 1.05 m above floor level. Replace faulty ironmongery.
Toughened glass failure	• Nickel sulphide inclusion.	Replace affected panes with good quality toughened or laminated glass.
Warping or cracking of uPVC framing or sashes	• Poor fit • Movement in masonry around opening/s.	Replace defective unit/s.

Note: attention is also drawn to the FENSA self assessment scheme administered by the Glass & Glazing Federation, accessible through the FENSA website (see Appendix).

Difficulties may be experienced in replacing frames dimensioned in imperial units if only metric standard units are available. In such cases it is more appropriate to replace with bespoke units. On external work particular care should be taken to re-establish waterproofing systems.

5.3.21 Slip resistance of floors

The HSE have investigated hundreds of accidents caused by people falling on defective or sub-standard floor surfaces. In an effort to standardise an approach to the prevention of this type of accident CIRIA have carried out a research programme in which a number of factors have been considered including:

- the flooring material and its roughness
- contamination of the surface
- the cleaning regime
- the footwear being worn
- environmental factors
- human factors affecting behaviour

The results of this research have been compiled in a soon to be published guide

5.3.22 Tenements

When the Scottish tenements were built 80–100 years ago, structural engineering skills were not applied to domestic architecture. The buildings were erected on well tried solutions and rule of thumb methods. It is doubtful if any stress calculations were done at all, for at best any structural problems were on the level of early editions of *Mitchells Building Construction* or Charles Gourley's *Elementary Building Construction for Scottish Students*. Nevertheless, they met the demand at that time for cheaply erected buildings using local materials, built to standards accepted at the time and with no real thought for their life span (see Fig. 5.34).

Lack of understanding has led to faults that include:

- Inadequate foundations
- Inadequate tie action between components

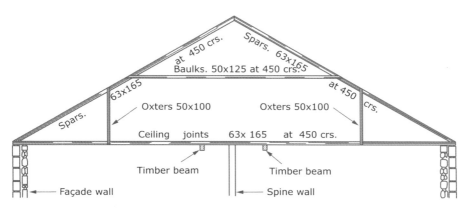

Figure 5.34 Typical box and oxter truss (Courtesy Prof. I A MacLeod).

- Lack of horizontal restraint
- Poor workmanship
- Deterioration in masonry due to damp
- Rot in timbers
- Rusting of steelwork.

5.3.23 Heritage and ecclesiastical buildings (including burial practices)

Practitioners should consult English Heritage (EH) and other similar authorities for advice on good practice in dealing with such structures but should be aware that in such work it is often a requirement to replicate contemporary materials and practices.

A full dissertation of ecclesiastical buildings is beyond the scope of this book. However useful material can be found in Jones (1965).

Until sometime in the 14th century it was accepted practice to bury human bodies within English churches. Attention is drawn to the need for careful investigation of areas within and nearby ancient churches to discover such burials before proceeding with new construction work. Modern techniques such as infrared spectroscopy and gas chromatography-mass spectrometry are available to investigate soil samples containing human remains and should be utilised if the presence of human burial is suspected. Several NDT methods are available for the location of buried objects (see Tomsett 2010).

5.3.24 Bridges

These may be constructed in concrete, steel, cast and wrought iron, timber, plastics or a combination of these.

5.3.24.1 Types of bridges

Suspension
- The Golden Gate Bridge, San Francisco, USA, opened in 1937 and is currently being retrofitted with additional measures to combat future earthquakes. These include strengthening saddle/cable connections, strengthening tower bases with additional steel plates and angles, confinement of concrete pier tops with prestressed steel tendons, installation of dampers at several locations along the deck, strengthening of pylons by internal reinforcement and strengthening of cable anchorages by internal reinforcement to housings.
- Clifton Bridge, 1836–1864, designed by IK Brunel with a clear span of 702 ft (see Fig. 7.11 in Chapter 7).
- Humber Bridge, 1978.

- Forth Road Bridge, 1964.
- Severn Bridge, 1966.

Simply supported or continuous beam
- A prestressed concrete bridge at Staples Corner on the A40 in London (concrete), recently repaired following terrorist action.
- Milford Haven (now Cleddau Bridge) (steel box girder), Pembrokeshire, Wales.

Cantilever
- Forth Rail Bridge, Scotland opened in 1889 and constantly being re-painted to repulse corrosion of structural members. One of the very early, major, steel structures.

Cable-stayed
- Oresund Bridge linking Denmark with Sweden.
- West Gate (Yarra) Bridge, Victoria, Australia.
- Boyne Bridge, Ireland (see Fig. 5.35).

Figure 5.35 Boyne Bridge in Ireland – an example of a cable-stayed bridge.

Moveable
- Runcorn–Widnes Transporter Bridge, Cheshire, England.
- Newport Transporter Bridge, South Wales (see Chapter 7 for details of major refurbishment).

Floating (or pontoon) bridges
- Lake Washington, Seattle, USA.

Arch
There are two main types of arch bridge, the steel structure of, for example, the Sydney Harbour Bridge and short span arches of brick, block or stone of which

Figure 5.36 Early iron arch bridge, Coalbrookdale, Shropshire, 1779.

there are possibly 60,000 examples in the UK. The former are prone to the usual problems of steel corrosion. The engineering profession is undecided as to the optimum way of analysing masonry arches. The traditional analytical tool is the 'Modified Mexe' method but a strong contender involves the use of 'ARCHIE', a computer programme developed by Professor William Harvey (bill@obvis.com). Defects in masonry arches include movement/settlement of abutments causing arch distortion; loose or missing bricks or blocks; poor quality and/or displacement of in-fill material and traffic or other damage to parapets (see Fig. 5. 37).

In a research programme featuring Belfast University and TRL a system for constructing short span arches made with engineered precast concrete blocks without the use of centring has been completed. The ARCHIE computer programme has been used to analyse the stability of the construction. This work may well see the re-emergence of short span masonry arches.

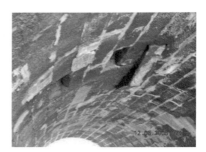

Figure 5.37 Arch bridge: note missing blocks (Courtesy Dr Bill Harvey).

Tied arch

Tyne Crossing, 1849. Designed by Robert Stephenson, the principal structures of which are the 38 m span cast iron arches. This bridge is the subject of much inspection, repair and load control to safeguard its future.

Bascule

- Tower Bridge, London opened in 1894.

Temporary

The best known UK temporary bridging is the eponymous prefabricated system designed during the Second World War by Sir Donald Bailey. Bailey bridges were built up from prefabricated units, the largest of which (the 10 ft × 5 ft, 570 lb panel) could be lifted by a team of six men (see Fig. 5.38). The steel used was of weldable quality; connections were made using simple steel panel pins. The system provided a clear road width of 10 ft 9 in. Live loads from Class 9 to Class 70 could be carried and bridges could be clear spanning or floated on pontoons. Bridges up to 200 ft clear spans could be provided. The simplest, lightest bridge consisted of single panels one storey high and was known as a 'single single'. The heaviest bridges consisted of triple panels two stories high and were known as triple doubles.

Practitioners examining existing structures should be aware that American and Italian versions exist that are not an exact match of the UK version.

Most such bridges were designed to carry battle tanks within the weight range 40–60 tonnes.

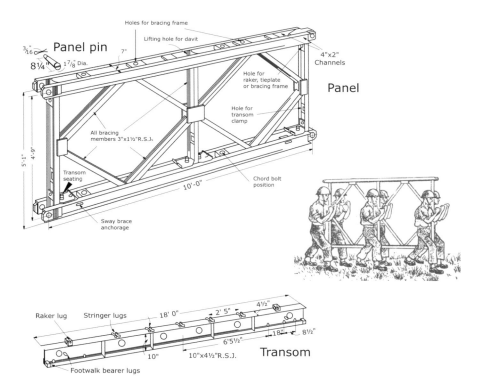

Figure 5.38 Bailey Bridge typical panel (redrawn from HMSO 1956).

- The original Bailey bridge has now been developed by Mabey and Johnson into the Super Bailey bridge in which the shear strength of the original panels has been increased by 33%. This enables double truss bridges to be used in situations that originally required triple trusses.
- In 1962 the Medium Girder bridge made from high strength welded aluminium was developed.
- It is interesting that strictly neutral Sweden is developing a new range of aluminium military bridges under the designation of Kb 71, capable of supporting a 69-tonne Leopard tank.

Bridge analysis and testing

Bridges under the jurisdiction of the Highways Agency are regularly inspected for defects. The following frequencies apply:

- Principal inspections: six-yearly. The purpose of this is to make a close examination of all inspectable parts of the structure, in particular to assess those parts where radical changes have been made to elements of the original structure.
- General inspections: two-yearly. The purpose of this is to inspect representative parts of a structure and assess defects.
- Superficial inspections: annually. The purpose of this is to report fairly obvious deficiencies which might, if ignored, lead to traffic accidents or high maintenance or repair cost.
- Special inspections: as required. The purpose of this is to make a close examination of any particular area of a structure whose performance is giving concern. The recommendation is to check cast iron structures, weight restricted structures and structures scheduled to carry abnormal loads at intervals not exceeding six months.

Although this is essentially HA advice it is relevant to and used by other authorities such as LUL.

As bridge engineers throughout the world are aware, situations arise where a theoretical analysis suggests inadequacies in structures that are apparently performing satisfactorily. Canadian and American engineers have addressed this problem by reviewing a series of tests (see Bibliography and further references). Baidir Bakht in his paper *Actual versus assumed behaviour of girder bridges* (1988) provides an excellent introduction to the phenomenon.

5.3.24.2 Problems with bridges

Scour

A frequent problem with bridges over water is the damage caused to piers and foundations by scour from fast flowing currents. Damage repair is difficult because of the need to carry out work below water level within a coffer dam. Useful advice on techniques for underwater inspection can be found in the IStructE document *Inspection of underwater structures* (2001).

Bridge bashing

This term is usually applied to bridges which have been damaged by collision with road or marine vehicles, for example a low bridge struck by a double-decker bus or the pier of a bridge over a river struck by a passing dredger (see Fig. 5.39). The Highways Agency is sacrificing lengths of hard shoulder either side of motorway bridges to install abutment and pier protection structures to avoid such damage. In Germany there have been reported cases of railway bridges being damaged by derailed trains. Footbridges across main roads have been demolished by vehicles carrying construction plant.

In all of these cases it is imperative that in addition to repair, consideration is given to better protection for the structure in future use.

Figure 5.39 This bridge – clearance 3.88 m (12'9") has been 'bashed' several times, at least once by a double-decker bus of height 14'6" (Courtesy Maureen Doran).

Alkali-silica reaction

See Section 5.2.1.10.

High winds

Frequently, many bridges are temporarily closed or have speed restrictions imposed to prevent the overturning of vehicles in crosswinds. Wind shielding is sometimes provided along the sides of bridges to increase the wind speed at which bridge closure becomes necessary. However, the barriers can cause significant increase in the lateral wind load imposed on the deck and may also affect the aerodynamic stability. Retrofitting of wind shielding may be economic where the cost of closure disruption is excessive. 18 m/s (40 mph) is the wind speed beyond which closure is considered by bridge authorities.

Wind shielding may be applied to the whole length or part of a bridge and may be solid or permeable. The design of these shields may be assisted by the results of wind tunnel analysis.

Durability of exposed members

There is a growing practice to protect exposed steelwork by encasements manufactured from plastics. In concrete bridge structures susceptible to ASR it is beneficial to increase the protection of the structure from water by improving drainage arrangements (see also comments in this chapter listed under concrete).

Insufficient reinforcement

Checks on reinforced concrete bridges have often revealed that the design reinforcement is insufficient for the increased load the bridge is now required to carry (or perhaps was under-designed in the original concept). Work by Professor Leslie Clark at Birmingham University has indicated that lack of shear strength adjacent to supports is perhaps a greater problem than lack of reinforcement to resist bending. Such a deficiency usually may be rectifiable by adding pre-stress where possible or building haunches in concrete or concrete-encased structural steel.

In cases where bending strength needs to be enhanced plate bonding using steel plates bonded to the parent concrete by epoxy resin may be considered. Alternative plate materials may be considered such as carbon fibre, kevlar and others which are not susceptible to corrosion.

Corrosion of pre-stressing cables

See Section 5.2.1.2.

Corrosion of suspension bridge cables

These cables may need periodic replacement due to corrosion. It is worth noting that at the time of writing the cables on the Forth Road bridge were being subjected to acoustic monitoring because corrosion was suspected. One report suggests that the bridge may have to close in 2013 due to corrosion. Also, corrosion has been reported on the suspension cables of the Severn Bridge where some cables have disintegrated due to water penetration either during construction or through cracks in paintwork during operation. It is planned to fit dehumidifying systems using warm dry air injection to extend the life of cables. At the time of writing this equipment was being installed.

Half joint corrosion

Post-Second World War, many bridges were constructed in reinforced concrete. Simply supported and cantilevered sections were often detailed with half joints which were vulnerable to corrosion due the ingress of de-icing salts (chloride). As a result, this type of joint has been largely discontinued. Where present in older designs considerable work has often been required to rebuild or replace the joint.

A recent example of this resulted in a section of a road bridge in Montreal, Canada collapsing with the resulting death of five people.

5.3.25 Tunnels

Tunnels are constructed by one of the following methods:

- Boring: This may be by hand, or more usually, carried out using a tunnel boring machine (TBM). In larger tunnels this will be a highly sophisticated mechanism which is capable of handling the spoil.
- Cut and cover: Usually employed for tunnels at shallow level, in which a trench is dug, the tunnel constructed and the soil then backfilled over the tunnel lining.
- Jacking: In certain specialised situations precast concrete sections are jacked through the ground from a jacking pit. This system was, for example, used for part of the Boston Freeway in America beneath existing, live rail tracks with a considerable saving of expense by comparison with competing methods. Other applications have involved jacking tunnels through railway embankments.
- Sprayed concrete: Using what is known as the New Austrian Tunnel Method, a technique is often used in which a tunnel is bored and reinforcement is fixed to the tunnel surface and then coated with sprayed concrete. It is essential that the condition of the unlined tunnel is constantly monitored to detect any movement in the ground as soil is removed.

In suitable, competent rock, tunnels may be unlined. Many examples also still exist of brick-lined tunnels used as sewers.

Problems in tunnels usually relate to rock falls in unlined tunnels or to corrosion damage in liners. Liners may consist of cast iron or precast concrete rings fitted in sections and bolted together or, alternatively, formed by concrete sprayed on to pre-positioned steel reinforcement. Concrete tunnel liners may be repaired using sprayed concrete or by liner replacement. Badly damaged cast iron liners may need to be removed and replaced either with new liners (specially produced cast iron or precast concrete) or with sprayed concrete.

Practitioners involved in projects that require new foundations or modifications of buildings need to be aware of the presence of existing tunnels perhaps at shallow depth. Examples of these are canalised rivers such as the Tyburn and the Fleet in London. Also the tunnels of the former Royal Mail Underground Railway and the Pneumatic Despatch Company both now disused (see Stanway 2000, 2002).

5.3.26 Cladding

Cladding is usually regarded as a non-load-bearing weatherproofing of a building – a climatic overcoat. In load-bearing masonry construction the structure performs a dual role – that of support and cladding. The air-tightness of buildings has become increasingly important in recent years as the regulatory authorities attempt to reduce energy consumption and the associated greenhouse gas emissions. Building Regulations Part L (2006) continues this trend, by requiring an improvement in energy efficiency compared with earlier Regulations. Improvements in the order 23.5–28% are called for. Air-tightness is likely to be central to the drive to meet these new standards. It should be noted that all buildings larger in area than 500 m² will require mandatory leakage tests to demonstrate compliance.

In refurbishment work the inclusion of electricity producing solar panels (photo-voltaic cells) should be considered as part of the cladding system.

Practitioners considering air-tightness of the building envelope should ensure that all components fit properly together and do not separate under load.

Cladding may appear as one or more of the following materials:

- boards (timber or other synthetic materials)
- concrete (precast concrete panels, block-work etc)
- facing brickwork (including brick slips)
- faience and terra-cotta
- glass
- glass curtain walling
- glass fibre reinforced cement (GRC)
- glass reinforced plastic (GRP)
- metal sheeting
- render
- slates
- stone
- tiling and mosaics (including tile hanging)
- timber facing
- composite panels

Refurbishment may be necessary for a variety of reasons including:

- Total or partial replacement.
- Improvement in aesthetic appearance.
- Improvement in thermal insulation.
- To provide better protection from strong sunlight.
- Lack of watertightness.
- To allow for expansion and contraction by insertion of appropriate joints.
- To improve or repair support and tie-back arrangements. Regrettably there have been a number of cases involving a framed building built considerably out of plumb on which a contractor has attempted to construct vertical cladding. The result has been that part way up the building the cladding has lost all vertical support. Best practice demands that a frame should be surveyed before cladding is commenced. The survey will then indicate the

level of any inaccuracies so that allowances can be made in the cladding support system (see Fig. 5.40).

- Surface cleaning.

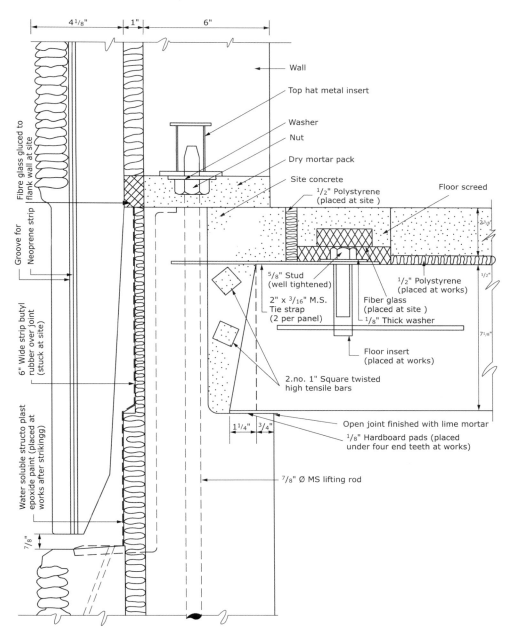

Figure 5.40 Typical best practice detail of cladding support.

5.3.26.1 Rain-screen

Allegedly developed in Scandinavia in the 1940s this concept is now widely used in Europe. A rainscreen system consists of an outer panel, a ventilated cavity (30–38 mm deep) and an inner leaf. Most rainwater is deflected off the outside face but any that penetrates is drained away. The system is pressure-equalised so precipitation is disinclined to be driven into the cavity.

5.3.26.2 Coated metal

In the early 1990s a rash of defects (more than 2000 cases) were identified in coated metal cladding. Causes of failure included:

- gloss change
- colour change
- crazing and flaking
- delamination
- corrosion at cut edge
- corrosion of base metal or random corrosion
- corrosion around fixings
- internal corrosion

BCSA *et al.* have recently drawn attention to the need to achieve air-tightness with metal cladding systems to comply with the updated Building Regulation Part L. They highlighted the fact that the requirement implies an increase in energy efficiency in the range 23.5–28%. Their practice note deals with the following topics:

- sizing of purlins and side rails
- choice of cladding system
- good site practice – purlins
- good site practice – cladding

5.3.26.3 Large concrete panels of solid and sandwich construction

Cases have been reported of deterioration of precast concrete panels which have led rise to doubts about their long term durability. The scope of this deterioration includes:

- Carbonation of concrete leading to corrosion of reinforcement.
- The presence of chlorides in widely varying proportions.
- Cracking at corners and arrises due to inadequate repair at time of manufacture or erection.
- Unrepaired cracks possibly caused during handling and erection.
- Exposed aggregate detachment which may be attributable to weathering.
- Crushing of panels due to elastic, creep or other thermal movements of the structure where provision of such movements has been inadequate.
- Delamination of layered concrete.

Remedial methods have been dictated by the severity of deterioration and the type of structure involved. These have varied between providing additional secure fixings, over cladding and, in extreme cases, removal and replacement of the existing cladding.

5.3.27 Asbestos

There are six main types of asbestos: *Chrysotile* (commonly known as white asbestos), *Amosite* (brown asbestos), *Crocidolite* (blue asbestos), *Anthophyllite*, *Actinolite* and *Tremolite*. White, brown and blue asbestos are the types most prevalent in construction. The use of blue and brown asbestos (the two most dangerous types) was banned in 1985; white asbestos was banned in 1999 (except for a small number of specialised uses). The asbestos content of materials varies from 85% (e.g. lagging, often brown or blue) to 10–15% (e.g. for asbestos-cement products; often white). Several other minerals are classed as asbestos but have not been commercially used.

Exposure to asbestos fibres may seriously damage the lungs and other organs resulting in diseases such as asbestosis, lung cancer and mesothelioma.

Many existing buildings contain these materials and laws now exist to control their encapsulation or removal (see Chapter 6). The following list gives an indication of possible uses of asbestos in buildings:

- Sprayed asbestos and asbestos loose packing – generally used as fire breaks in ceiling voids.
- Moulded or preformed lagging – generally used in thermal insulation of pipes and boilers.
- Fire-proof gasketting to boilers and fires.
- Sprayed asbestos – generally used as fire protection in ducts, firebreaks, panels, partitions, soffit boards, ceiling panels and around structural steelwork.
- Insulating boards used for fire protection, thermal insulation, partitioning and ducts.
- Ceiling tiles.
- Millboard, paper and paper products used for the insulation of electrical equipment. Asbestos paper has also been used as a fire-proof facing on wood fibreboard.
- Asbestos cement products, which can be fully or semi-compressed into flat or corrugated sheets. Corrugated sheets are largely used as roofing and wall cladding. Other asbestos cement products include gutters, rainwater pipes and water tanks.
- Some textured coatings.

- Bitumen roofing material.
- Vinyl or thermoplastic floor tiles.
- Artex paint often used to decorate ceilings.

If asbestos is suspected then a careful programme of inspection and testing leading to a report with remedial recommendations should be put in hand. Such investigations should only be carried out by those specially qualified so to do.

By law, if asbestos is present, it should either be removed or encapsulated. Encapsulated asbestos should be clearly marked by labels bearing the acronym ACM (Asbestos-Containing Materials). Removal will require the employment of registered contractors skilled in this complex type of operation. (There has, however, been a recent slight relaxation concerning the removal of Artex materials.) At the completion of the process a certificate must be obtained from an independent laboratory to verify that the concentration of airborne fibres is at or below statutory levels.

It is customary to record the findings in an Asbestos Register at the conclusion of a survey. This document can then be periodically updated to record any significant change to the presence or type of asbestos material in a building.

5.3.28 Japanese knotweed and other injurious weeds

Japanese knotwood was introduced to the UK in the 1800s as an ornamental plant and is now one of our most invasive and destructive species. The roots of this plant have been reported to pass through the joints of brickwork and to break masonry apart. It has become a serious menace to new construction. The situation has been exacerbated by careless fly tipping by those wishing to rid themselves of the plant. The plant can grow to a height of 3 m, overshadowing and suppressing the growth of more desirable flora. Treatment should be radical: removal by biological or other methods.

It is now an offence under the *Wildlife and Country Act 1981* to encourage the growth of this plant and also giant hogweed. Defra has estimated the cost of controlling this menace at many millions of pounds. Although eradication methods are still emerging as a result of research, the most common are herbicide spraying or removal to a regulated land-fill site where it is buried beneath a geotextile membrane layer surmounted by 5 m of clean fill material.

Further injurious weeds covered by the *Weeds Act 1959* include spear thistle, creeping or field thistle, curled dock, broad leaved dock and common ragwort.

5.3.29 Service installations

Service installations are likely to need alteration, repair or replacement during the refurbishment of a building. The following services should be considered:

- gas
- water
- electrical

- drainage
- lift installations
- data connections, including telephones, television and wireless connection of sound, vision and data
- district heating and combined heat and power systems
- mechanical and natural ventilation
- heating, cooling and humidity control
- techniques for reducing service loads

The reasons for changes to services and equipment will be many and varied:

- new technical requirements
- change of use of the building
- unsuitable or out of date services
- services required in new locations
- compliance with current legislation

The location and types of services available both inside and outside a building need to be identified. Decisions need to be taken on whether to alter or replace the services to suit new requirements. The services inside a building and those with connections to outside supplies need to be considered at an early stage of the design process. New use of a building could increase the load on service connections leading to replacement of incoming mains.

Services need to be considered alongside other aspects of the built environment including:

- the condition and construction of the building fabric
- thermal performance of the fabric
- mechanical and natural ventilation
- prevention of condensation
- cost in use, energy conservation and recycling
- maintenance

5.3.29.1 The condition and construction of the building

Knowledge of any existing building is essential at the start of a project. Check the structural stability and establish what alterations can be made to accommodate services. Any potentially harmful materials need to be identified and procedures put in place for safe methods of working with them in place or the removal of the materials. The following should be considered:

- Structural stability. Establish if it safe to work in the building and what changes can be made to structural elements to accommodate services.
- Hazardous materials containing asbestos. These could include pipe work insulation, seals in heating equipment, wall and roof cladding, wall and ceiling lining, vinyl and other floor tiling, decorative finishes.

- Potentially harmful materials such as dust, silica, fungal growths and fumes in confined spaces.
- Flammable materials. Gas in pipe work, petrol and oil storage containers and pipe work.
- Contamination from water supplies, cooling systems and drains.
- Animal infestation and infectious diseases.

5.3.29.2 Services connected to outside the building

The availability, capacity and location of the existing services will have an influence on the design of a project. The following should be considered:

- The availability and capacity of the service in the area. Establish if gas, water, drains, electrical and data connections are available and adequate to the needs of the project. If necessary find out if upgrading mains supplies is possible and the cost and the time it would take. If new plant is required to achieve the capacity (such as an electrical transformer) consider the location of the equipment, buildings to accommodate plant and the granting of leases to the service provider.
- The capacity of services into the building. Establish if the capacity of connections are adequate. Check the capacity of gas, water and drainage services against any new requirements. For electrical supplies check the capacity and whether the supply is single or three phase. For data services check the type and capacity of services (such as fibre optic connections, cable services and wireless reception).
- The location of the service on entry into the building. Connecting gas, water and electrical installation to an existing point of entry might be possible. Connections to existing drains might be limited by the depth of the system and this will have an effect on the design options available.
- Find out if the services are shared with other properties or pass through other properties. The legal rights and obligations should be established.
- Restriction could be placed on building over shared drains or services.

5.3.29.3 Existing and adapted services within a building

Existing services might be suitable to be retained or adapted. Check the existing service installations including the following:

- Suitability for the project.
- Compliance with current requirements including:
- The Building Regulations.
- Gas Safe Register guidance (previously CORGI – Council of Registered Gas Installers).
- IEE regulations (Institution of Electrical Engineers).
- Health and safety regulations could affect or determine if plant can be retained. This would have an effect on all services including safe access to

plant and equipment, lift installations, water installations (and health risks such as legionella), boilers and ventilation.

- Planning and Listed Building legislation.
- The Disability Discrimination Act could affect the options available.
- If adaptation is required check if suitable parts and materials are available. Cast iron pipe work is available in a more limited range than in the past. Changes from the imperial to metric system have limited the options on older installations. Availability and the maintenance costs of old equipment might be prohibitive.
- Any retained system might be affected by changes in other parts of the installation. For instance, lime scale in water installations might be loosened by vibration during work and cause blockages in ball valves and water outlets.
- On site storage of energy would need to comply with current legislation including gas, oil and petrol containment.

5.3.29.4 *New services within a building*

Installing new services in an existing building requires the identification of where the services can be run, routes for cables, water pipes, heating installations, flues, ventilation ducts and data installation. The following should be done:

- Coordinate the various services with the structure.
- Coordinate services with any upgrading of the fabric including thermal, fire, security and sound installations.
- Identify planning restrictions and Listed Building requirements that could limit the options available for services on the outside of a building. Flues and vents might be limited to rear elevations or to roof level. Restrictions could also apply internally.
- For lift installation or renewal check the adequacy of electrical supply. The installation of an evacuation lift would require compliance with more onerous supply and cabling conditions.
- If there are existing cavities within a structure consider using them for services. Check that access, maintenance, fire and sound insulation are not compromised.

Room heights will influence whether lowered ceilings can be installed to contain services. Room heights and the retention of existing stairs will influence whether raised floors can be installed to contain services such as data and power.

5.3.29.5 *Techniques for reducing service loads*

Reducing the level of demand on services is an approach that needs to be considered at the earliest possible stage of a project. Avoiding an increase in demand on incoming services could lead to the retention of existing mains, lower operating costs and increased sustainability. Consider:

- Improved insulation of the fabric.
- Energy efficient equipment and controls.
- Design for and use of natural ventilation rather than cooling plant.
- Existing drainage capacity might restrict the quantity of storm water allowed to enter the drainage system. This could be achieved by retaining rainwater on site to allow gradual run off. Techniques to achieve this include soak-aways, storage ponds and green roofs that retain water.
- Reductions in water usage could include harvesting rainwater to use in non-potable application such as flushing toilets.
- Electrical requirements could be reduced by generation on site from wind turbines and photovoltaic cells.
- Solar water heating.
- Use of ground source heat pumps.
- Small-scale hydro-electric generation.
- Biomass boilers using forest or other products.

5.3.30 Underground services

Piped underground services providing for drainage, gas, water, sewage may be constructed using a variety of materials including:

- vitrified clay (to BS EN 295 or BS 65)
- concrete (plain, reinforced or prestressed to BS EN 1916 and 1917)
- grey iron (to BS EN 877)
- ductile iron (to BS EN 545; BS EN 545;BS EN 598 and BS EN 969)
- glass-fibre reinforced plastics (GRP) (to BS 5480)
- steel
- unplasticised PVC (PVC-U) (to BS EN 1401)
- polyethelene (to BS 6572; BS 6437 or BS 3284)
- polypropylene (to BS EN 1852-1)

Electrical, telephonic and cable services are to be found as sheathed cables. They are often protected by ceramic tiles to resist local damage.

Older sewers such as those constructed in London under the supervision of Joseph Bazalgette (1819-1891) were constructed in brick and have survived extremely well.

For those wishing to repair or extend underground systems, a major difficulty lies in locating existing services due to lack of records. This situation has been exacerbated by the decline in centrally held records in the hands of District Surveyors and Local Authority Chief Engineers. The National Joint Utilities Group (NJUG) may be able to assist. If these records are unavailable then recourse must be made to careful excavation or the use of sub-surface radar and other NDT locating methods. See Fig 5.41 for a recommended arrangement of services.

Until the recent past it has been customary for foul and storm-water drainage to be accommodated in the same sewer system. However in more recent times best

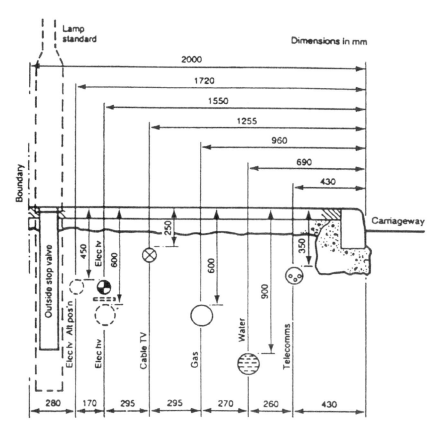

Figure 5.41 Recommended arrangement of mains in a 2 m footway including cable TV duct.

practice indicates that the two should be run in separate systems. With predicted water shortages there is a move to provide housing and other accommodation with a grey water system for flushing toilets and other similar activities.

Given sufficient pipe diameter and reasonably straight runs then closed circuit television (CCTV) techniques will assist in checking the condition of drain lines.

Techniques employed in renovating some services include:

- Insertion of new plastic liners to existing pipe-work. This technique has been extensively used by British Gas in updating main supply lines. The water industry has a good record of lining existing pipelines with GRP liners.
- Connecting or inserting metric sized piping to existing imperial sized mains. Coupling units are available to assist in this task.
- Partial replacement of brickwork in old brick sewers and constructing new manholes using sectionalised precast concrete units.

Bibliography and further reference

Note:

(1) All IABSE British Group reports are available via the Institution of Structural Engineers (IStructE) Library. For all BCA and C&CA reports check with The Concrete Centre, Camberley, Surrey, GU17 9AB. www.concretecentre.com

(2) Nikolaus Pevsner has written a series of books under *The Buildings of England* series dealing with architecture in many parts of England. They form a useful starting point for those seeking construction dates and details of many buildings.

5.1 General

ACE. 1969. *Emley Moor –ice loading design,* ACE, London.

Allinson, K. 2006. *London's Contemporary Architecture,* Architectural Press, Oxford.

Bate, S.C.C. 1974. *Report on the failure of roof beams at Sir John Cass's Foundation and Redcoat School, Stepney, BRE CP.58/74,* BRE, Garston, UK.

Beckmann, P. and Bowles, R. 2004. *Structural aspects of building conservation,* 2nd edn., Elsevier, Oxford.

BRE. 1974. *Floor loadings in office buildings – the results of a survey,* CP 3/71, BRE, Garston, UK.

BRE. 1975. Failure patterns and implications, *Digest 176,* BRE, Garston, UK.

BRE. 1982. Common defects in low-rise traditional housing, *Digest 268,* BRE, Garston, UK.

BRE. 1991. Why do buildings crack?, *Digest 361,* BRE, Garston, UK.
Additional information is to be found in the BRE series *Defect Action Sheets.* DAS 1–99. Although issued in the period 1982–1987 many of these defects are still to be found in housing today.

BS 6399-1: 1984. *Design loadings for buildings Part 1. CP for dead and imposed loads,* British Standards Institution, London.

Campbell, P. (Ed.) 1997.

Campbell, P. (Ed.) 2001. *Learning from construction failures,* Whittles Publishing, Scotland. Includes details of the collapse of the walkway at the Hyatt Regency Hotel, Kansas City, Missouri.

Carillion, 2001. *Defects in buildings,* HMSO, London.

Chapman, J.C. 1998. Collapse of the Ramsgate Walkway, *The Structural Engineer,* 76(1) IStructE, London.

Clark, L.A., Shammas-Toma, M.G.K. , Seymour, D.E. , Pallett, P.F. and Marsh, B.K. 1997. How Can We Get the Cover We Need? *The Structural Engineer,* 75(17), IStructE, London.

Constable, A. and Lamont, C. 2006. *Building defects,* RICS Books, London.
This book deals with the legal aspects of defects including: what is a defect; defects liability; temporary disconformity; claims for defective work in tort; the Defective Premises Act 1972; the surveyors duty to identify defects; defects and limitation periods; defects and the project team; and remedies.

Coxon, R.E. 1986. *Failure of the Carsington embankment,* HMSO, London.

Cullen, W.D. 1990, *The public enquiry into the Piper Alpha disaster.* 2vols, Department of Energy, HMSO, London.

DOE/SDD/WO. 1971. *Inquiry into the basis of design and method of erection of steel box girder bridges*, HMSO, London.

Doran, D.K. (Ed.) 1991. *Eminent Civil Engineers: their 20th century life and times*, Whittles Publishing, Scotland.

English Heritage. 1994. *Office floor loading in historic buildings*, English Heritage, London.

Elliott, C.L., Mays, G.C. and Smith, P. D. 1992. The protection of buildings against terrorism and disorder, *Proceedings of the ICE, Structures and buildings*, 94(3), ICE, Thomas Telford, London.

Foster, J.S. 1963. *Mitchell's advanced building construction*, 17th edn., Batsford, London. This book was originally produced in 1893. Earlier editions are still available from reputable libraries and can be of considerable assistance in discovery of early types of construction.

Harvey, W. 1999. Learning from failures. In: Liddell, I. (Ed.) *Learning from Engineering History*, Proceedings of the 1999 Henderson Colloquium, organised by the British Group of the IABSE, London.

Home Office. 1989. *The Hillsborough stadium disaster.* The Taylor Report, HMSO, London.

HSE.1985. *The Abbeystead Explosion*, Dd 715226 C60 2/85, HSE, London.

HSE. 2000. *The collapse of NATM tunnels at Heathrow Airport (Oct 1994)*, HSE, London.

Ingham, J. 2009. Forensic engineering of fire-damaged structures, *Proceedings of ICE, Civil Engineering*, 160(3), Thomas Telford, London.

IStructE. 1966. *Industrialised building and the structural engineer*, IStructE, London.

IStructE. 1991. *Surveys and inspections of buildings and similar structures*, IStructE, London.

IStructE. 1996. *Appraisal of existing structures*, 2nd edn., IStructE, London. New edition is in preparation.

IStructE. 2002. *Safety in tall buildings*, IStructE, London. Report dealing with the terrorist action on the WTC twin towers in New York, 11 September 2001.

IStructE. 2002. *Design recommendations for multi-storey and underground car park*,. IStructE, London. Report produced in response to the terrorist attack on the World Trade Centre, New York, 11 September 2001.

IStructE. 2003. *Introduction to fire safety engineering*, IStructE, London.

Jennings, A. 2006. Using disasters as a learning tool in higher education, *The Structural Engineer*, 84(15), IStructE, London.

Mainstone, R.J. and Butlin, R.N. 1976. *Report on an explosion at Mersey House, Bootle, Lancs*, BRE CP 34/76, BRE, Garston, UK.

Mann, A. 2006. Construction safety; an agenda for the profession, *The Structural Engineer*, 84(15), IStructE, London.

Matousek, M. 1977. Outcomings of a survey on 800 construction failures. In: Sandberg, A. (Ed.) *Inspection and quality control*, Proceedings of the 1977 Henderson Colloquium, organised by the British Group of the IABSE, London.

Menzies, J.B. and Grainger, G.D. 1976. *Report on the collapse of the Sports Hall at Rock ferry Comprehensive School Birkenhead*, BRE CP 69/76, BRE, Garston, UK.

Middleton, G.A.T. *c.*1900, *Modern buildings, their planning, construction and equipment*, Caxton Publishing, London.

Ministry of Power. 1967. *Report of the enquiry into the causes of the accident to the drilling rig Sea Gem*, HMSO, London.

Ministry of Housing and Local Government. 1968. *Collapse of flats at Ronan Point, Canning Town,* HMSO, London.

NBA. 1987. *Common building defects: diagnosis and remedy,* National Building Agency, Longman, Harlow,

Paczac, M.G., Duvall, P.E. and Cosby, J. 2005. *Blast-resistant design for buildings,* Go-Structural.com. Available at: www.gostructural.com/article.asp?id=279. Accessed 01 June 2009.

Paterson, A. 1984. Presidential address: The Structural Engineer in Context, *The Structural Engineer,* 62(11), IStructE, London.

Rock, R.A. and MacMillan, I.R. (Eds.) 2005. *Victorian House Manual,* Haynes, Yeovil.

Rock, R.A.and MacMillan, I.R. (Eds.) 2005. *1930s House Manual,* Haynes, Yeovil.

Royal Commission (Australia). 1971. *Report of Royal Commission: Failure of West Gate Bridge,* RC (Australia), Govt Printer, Melbourne.

Rushton, T. 2006. *Investigating hazardous and deleterious building materials,* RICS Books, London.

SCOSS. 1986. *Seventh report, giving brief details of the collapse of the YNYSYGWAS bridge,* IStructE, London.

SDD. 1972. *Clarkson Toll: fatal accident enquiry, Edinburgh,* SDD, Edinburgh.

Silcock, A. and Hinkley, P.L. 1974. *Report on the spread of fire at Summerland,* BRE CP 74/74, BRE, Garston, UK.

Smith, D. (Ed.) 2001. *Civil engineering heritage: London and the Thames Valley,* Thomas Telford, London.

Sriskandan, K. 1984. Ways of preventing failure. In: Sandberg, A. (Ed.) *Liability,* Proceedings of the 1984 Henderson Colloquium, organised by the British Group of the IABSE, London.

Wearne, P. 1999, *Why buildings fall down,* Channel 4 Books, London.

Wood, J.G.M. 1999. Communicating lessons from failures: concisely. In: Liddell, I. (Ed.) *Learning from Engineering History,* Proceedings of the 1999 Henderson Colloquium, organised by the British Group of the IABSE, London.

Wood, J.G.M. 1995. Silos: evolution by failure. In: Simpson, B. (Ed.) *Containment Structures,* Proceedings of the 1995 Henderson Colloquium, British Group of the IABSE, London.

5.1.3 Relocating structures

Charge, J. 1972. The raising of the Old Wellington Inn and Sinclair's Oyster Bar, *The Structural Engineer,* 50(12), IStructE, London.

Olsen, K.A. 1958. The re-siting of structures, *Proceedings of the 50th Anniversary Conference 1958,* IStructE, London.

Pryke, J. 1987. Raising and moving buildings. In: Sandberg, A. (Ed.) *Rehabilitation and renovation,* Proceedings of the 1987 Henderson Colloquium, British Group of the IABSE, London.

5.2 Basic materials

Sheehan, T. 1997. *Good practice in the selection of construction materials,* BCO (in conjunction with ARUP and BPF), Reading.

5.2.1 Concrete

5.2.1.1 General

BCA. 1999. *Concrete through the ages,* British Cement Association, Surrey.

BCA *et al.* 2000. *Improving concrete frame construction,* British Cement Association, Surrey.

BCA *et al.* 2000. *Concreting for improved speed and efficiency,* British Cement Association, Surrey.

BCA *et al.* 2000. *Early age strength assessment of concrete on site,* British Cement Association, Surrey.

BCA *et al.* 2000. *Improving rebar information and supply,* British Cement Association, Surrey.

BCA *et al.* 2000. *Rationalisation of flat slab reinforcement,* British Cement Association, Surrey.

BCA *et al.* 2001. *Early striking and improved back-propping for efficient flat slab construction,* British Cement Association, Surrey.

BCA, *et al.* 2001. *Flat slabs for efficient concrete construction,* British Cement Association, Surrey.

BCA and Concrete Centre. 2007. *CO_2 uptake from the re-carbonation of concrete,* Innovation & Research Focus.

Beckmann, P. and Bowles, R. 2004. *Structural aspects of building conservation,* 2nd edn., Elsevier, Oxford. (Chapter 7.)

BRE. 1993. *Concrete: cracking and corrosion of reinforcement,* Digest 389, BRE, Garston, UK.

BRE. 1999. *Concrete corrosion – a £550m- a-year problem,* Research Focus, ICE, Garston, UK.

Browne, R.D. 1978. Assessment of the strength in concrete structures and identification of corrosion in reinforcement. In: *Inspection and maintenance,* Proceedings of the 1978 Henderson Colloquium, organised by the British Group of the IABSE, London.

Bungey, J.H. 1983. The role of NDT for concrete structures. In: Somerville, G. (Ed.) *Instrumentation of structures,* Proceedings of the 1983 Henderson Colloquium, organised by the British Group of the IABSE, London.

Chapple, P.G. and Doran, D.K. 1978. Investigation assessment and repair of reinforced concrete framed building damaged by corrosion of reinforcement. In: *Inspection and maintenance,* Proceedings of the 1978 Henderson Colloquium, organised by the IABSE, London.

CIRIA. 1984. *Spalling of concrete in fires,* Technical Note 118, CIRIA, London.

CIRIA. 1987. *Protection of reinforced concrete by surface treatments.* Technical Note 130, CIRIA, London.

CIRIA. 1992. *Testing concrete in structures,* Technical Note 143, CIRIA, London.

CIRIA. 2001. *Specifying, detailing and achieving cover to reinforcement,* Report C568, CIRIA, London.

CIRIA. 2007. *Guide to early thermal cracking,* Report R660, CIRIA, London.

Clark, L.A., Doran, D.K. and Lazarus, D. 2002. *Quality of in-situ concrete construction in the UK,* JCSA, Tokyo.

Clark, L.A., Shammas-Toma, M.G.K., Seymour, D.E. , Pallett, P.F. and Marsh, B.K. 1997. How can we get the cover we need?, *The Structural Engineer,* 75(17), IStructE, London.

CS *et al.* 1984. *Repair of concrete damaged by reinforcement corrosion,* Technical Report No. 26, CS, Surrey.

Deacon, R.C. 1973. *Watertight construction,* C&CA, Slough.

EMPA. 1964. *A concrete bridge destroyed to test it,* Swiss Federal Laboratories for Materials Testing and Research, Dübendorf. This publication describes a test on a 23 m span pre-stressed concrete bridge in Switzerland built in 1954–1955 but demolished in the early 1960s to accommodate a new highway.

FIP/CEB. 1978. *Report on methods of assessment of the fire resistance of concrete structural members,* FIP/CEB, Slough.

Forrester, U.A. *et al.* 1978. The identification of current special problems with concrete in structures. In: *Inspection and maintenance,* Proceedings of the 1978 Henderson Colloquium, organised by the IABSE, London.

Hewlett, P.C. 1981. The repair of concrete and the use of organic polymers to achieve the repair. In: Cusens, A. (Ed.) *Materials in structures,* Proceedings of the 1981 Henderson Colloquium, organised by the British Group of the IABSE, London.

Hurst, B.L. 1996. Concrete and the structural use of cements in England before 1890. *Proceedings of the ICE, Structures and Buildings,* 116(3), ICE, London.

ICE. 1996. Historic Concrete, *Proceedings of the ICE, Structures and Buildings,* ICE, London. A special issue of *Structures and Buildings* August/November, 1996. This set of papers was followed by the issue of a corrigenda issued in August/November 1996 with corrections to papers by Frank Newby and Michael Bussell.

Somerville, G. 1986. The design life of concrete structures. *The Structural Engineer,* 64(2), IStructE, London.

Stanley, C.C. 1979. *Highlights in the history of concrete,* C&CA, Slough.

Sutherland, R.J.M., Hume, D., Chrimes, M. (Eds.) 2001. *Historic concrete: background to appraisal,* Thomas Telford, London.

5.2.1.2 Prestressed concrete

Blake, L.S. (Ed.). 1989. *Civil engineer's reference book,* Butterworth, Oxford. (Chapter 12.)

CS. 1979. *Flat slabs in post-tensioned concrete with particular regard to the use of unbonded tendons – design recommendations,* Technical Report No. 17, CS, Surrey.

CS. 2002. *Durable post-tensioned concrete bridges,* 2nd edn., Technical Report No. 47, CS, Surrey.

Hollinghurst, E. 1999. Post-tensioned concrete and corrosion. In: Liddell, I. (Ed.) *Learning from engineering history,* Proceedings of the 1999 Henderson Colloquium, organised by the British Group of the IABSE, London. A report of a Concrete Society Working Party in collaboration with the Concrete Bridge Development Group.

Somerville, G. 1997. The performance in service of concrete bridges. In: Nethercot, D. (Ed.) *Structures for serviceability,* Proceedings of the 1997 Henderson Colloquium, organised by the British Group of the IABSE, London.

5.2.1.3 Glass-fibre reinforced cement [GRC]

Doran, D.K. (Ed.) 1992. *Construction materials reference book,* Butterworth-Heinemann, Oxford, Ch 20. (A second edition is in preparation.)

GRCA. 2006. *Design guide,* International Glassfibre Reinforced Concrete Association, Surrey.

5.2.1.4 Sprayed concrete

Doran, D.K. (Ed.) 1992. *Construction materials reference book,* Butterworth-Heinemann, Oxford, Ch 24. (A second edition is in preparation.)

Hewlett, P.C. (Ed.) 2003. *Lea's Chemistry of cement and concrete.* 4th edn., Butterworth-Heinemann, London, Ch 15.

IStructE. 2004. *Design and construction of deep basements including cut-and-cover structures,* IStructE, London, Appx E.

5.2.1.5 Reinforced autoclaved aerated concrete (RAAC)

Desai, S. 2002. *Reinforced autoclaved aerated concrete roof slabs,* Building Engineer, Northampton.

Desai, S. 2004. *Appreciation of risks in specifying and designing concrete structures,* Building Engineer, Northampton.

Matthews, S., Narayana, N. andGoodier, A. 2002. *Reinforced autoclaved aerated concrete panels: review of behaviour, and developments in assessment and design,* BRE Press, Garston, UK.

5.2.1.6 Mundic

BRE. 1992. *Taking care of 'mundic' concrete houses,* BRE, Garston, UK.

IStructE. 1988. *Mundic,* Interim Technical Guidance Note, IStructE, London.

RICS. 1997. *The 'Mundic' problem – a guidance note: recommended sampling, examination of suspect building materials in Cornwall and parts of Devon,* RICS, London.

5.2.1.7 Renders and plasters

BRE. 1976. *External rendered finishes,* Digest 196, BRE, Garston, UK.

Doran, D.K. (Ed.) 1992. *Construction materials reference book,* Butterworth-Heinemann, Oxford. (A second edition is in preparation.)

Noy, E.A. and Douglas, J. 2005. Building Surveys and Reports, Blackwell, Oxford.

5.2.1.8 Large panel systems (see cladding)

5.2.1.9 High Alumina Cement (HAC)

CS. 1997. *Calcium aluminate cements in construction,* The Concrete Centre, Slough.

5.2.1.10 Alkali-silica reaction [ASR]

BCA. 1988. *The diagnosis of alkali-silica reaction,* BCA, Slough.

BRE. 1997. *Alkali-silica reaction in concrete,* Digest 330, Pts 1, 2, 3 and 4, BRE, Garston, UK.

BRE. 2002. *Minimising the risk of alkali-silica reaction: alternative methods,* Information Paper 1/O2, BRE, Garston, UK.

CS. 1999. *Alkali-silica reaction: minimising the risk of damage to concrete,* Technical Report No: 30, CS, Surrey.

IStructE. 1999. *Structural effects of alkali-silica reaction: technical guidance on the appraisal of existing structures,* IStructE, London.

5.2.1.11 Deleterious aggregates

Chapple, P.G. and Doran, D.K. 1978. Investigation assessment and repair of reinforced concrete framed building damaged by corrosion of reinforcement. In: *Inspection and maintenance*, Proceedings of the 1978 Henderson Colloquium, organised by the IABSE, London.

Hewlett, P.C. (Ed.) 2003. *Lea's Chemistry of cement and concrete.* 4th edn., Butterworth-Heinemann, London.

Regan, P.E., Kennedy-Reid, I.L., Pullen, A.D. and Smith, D.A. 2005. The influence of aggregate type on the shear resistance of reinforced concrete, *The Structural Engineer*, 83(23), IStructE, London.

5.2.1.12 Carbonation

Yu, C.W. and Bull, J.W. 2006. *Durability of materials and structures: in building and civil engineering*, Whittles Publishing, Scotland.

5.2.1.13 Hydrogen embrittlement

Bentur, A. *et al.* 1997. *Steel corrosion in concrete*, E and FN Spon, London.

Bertolini, L., Elsener, B., Pedeferri, P. and Polder, R. 2004. *Corrosion of steel in concrete*, Wiley-VCH Verlag GmbH & Co. KGaA, Weinheim, Germany.

5.2.1.14 Rust staining [see text]

5.2.1.15 Acid and sulfate attack (including thaumasite)

DETR. 1999. *The Thaumasite form of sulfate attack*, Report of the Thaumasite Expert Group. DETR, London.

5.2.1.16 Woodwool formwork

BRE. 1978. *An investigation into the fire problems associated with wood wool permanent shuttering for concrete floors*, BRE, Garston, UK.

John Laing Research and Development Ltd. and WWSMA. 1975. *Investigation into the use of wood-wool as permanent shuttering.* First series report for the Wood Wool Slab Manufacturers Association, London.

5.2.1.17 Repair and strengthening

ACI/BRE/CS *et al.* 2003. *Concrete repair manual*, 2 vols., CS, Surrey.

BS 7973-1: 2001. *Spacers and chairs for steel reinforcement and their specification. Product performance requirements*, BSI, London.

BS 7973-2: 2001. *Spacers and chairs for steel reinforcement and their specification. Fixing and application of spacers and chairs and tying of reinforcement*, BSI, London.

CS. 1984. *Repair of concrete damaged by reinforcement corrosion*, Technical Report No. 26, CS, Surrey.

Hewlett, P.C. 1981. The repair of concrete and the use of organic polymers to achieve the repair. In: Cusens, A. (Ed.) *Materials in structures*, Proceedings of the 1981 Henderson Colloquium, organised by the British Group of the IABSE, London.

5.2.1.18 Cathodic protection

Note: All CPA documents are available **free** from their website: www.corrosionprevention.org.uk.

CPA. 1998. *Reinforced concrete: history, properties and durability*, Technical Note No. 1, CPA, Aldershot, UK.

CPA. 2002. *An introduction to electrochemical rehabilitation techniques, Technical Note No: 2*, CPA. Aldershot.

CPA. 2002. *Cathodic protection of steel in concrete. The international perspective, Technical Note No: 3*, CPA. Aldershot.

CPA. 2002. *Monitoring and maintenance of cathodic protection systems, Technical Note No: 4*, CPA. Aldershot.

CPA. 2002. *Corrosion mechanisms – an introduction to aqueous corrosion, Technical Note. No: 5*, CPA. Aldershot.

CPA. 2002. *The principles and practice of galvanic cathodic protection for reinforced concrete structures, Technical Note No:6*, CPA, Aldershot.

CPA. 2002. *Cathodic protection of early steel framed buildings, Technical Note No: 7*, CPA, Aldershot.

CPA. 2002. *Cathodic protection of steel in concrete – frequently asked questions, Technical Note: No: 8*, CPA, Aldershot.

CPA. 2004. *Electrochemical re-alkalisation of steel reinforced concrete – a state of the art report, Technical Note: No: 9*, CPA, Aldershot.

CPA. 2002. *Stray current, Technical Note: No: 10*, CPA, Aldershot.

HA. 2002. *Cathodic protection for use in reinforced concrete highway structures, BA 83/02*, HA, London.

IStructE. 1988. *Cathodic protection of concrete structures*, IStructE, London.

5.2.1.19 Desalination and re-alkalisation

Broomfield, J.P. 2004. *Electrochemical re-alkalisation of steel reinforced concrete*, CPA Technical Notes No.9, CPA, Aldershot.

Meitz, J. (Ed.) 1998. *Electrochemical rehabilitation methods for reinforced concrete*, Woodhead Publishing, Cambridge.

5.2.1.20 Coatings

Bassi, R, and Roy, S.K. (Eds.) 2002. *Handbook of coatings for concrete*, Whittles Publishing, Scotland.

Biczok, I. 1972. *Concrete corrosion, concrete protection*, Akademiai Kiado, Budapest.

CIRIA. 1987. *Protection of reinforced concrete by surface treatments*, Technical Note 130, CIRIA, London.

5.2.1.21 Industrialised building systems
Housing

BRE. 1989. *The structural condition of: Wimpey no-fines low rise dwellings*, BR153, BRE, Garston, UK.

Williams, A.W. *et al.* 1991. *The renovation of no-fines houses*, BR191, BRE, Garston, UK.
Note: This report is based on SSHA and Wimpey systems.

Other concrete systems

BRE. 2004. *Non-traditional houses: identifying non-traditional houses in the UK 1918–1975,* Report 469, BRE, Garston, UK.

BRE. 1970. *The comprehensive industrialised building systems annual,* BRE, Garston, UK.

5.2.2 Masonry

5.2.2.1 General

Bensalem, A. *et al.* 1997. Non-destructive evaluation of the dynamic response of a brickwork arch, *Proceedings of ICE, Structures and Building,* 122(1), ICE, London.

BRE. 2004. *Structural fire engineering design: materials behaviour–masonry,* BRE, Garston, UK.

BRE. 1973. *Long term expansion of test brick, CP 16/73,* BRE, Garston, UK.

BRE.1988. *Inserting wall ties in existing construction,* Digest 329, BRE, Garston, UK.

BRE. 1997. *Repairing damage to brick and block walls, GR 3,* BRE, Garston, UK.

BRE. 1996. *Replacing masonry ties, GR4,* BRE, Garston, UK.

BRE. 1983. The selection of natural building stone, *Digest 269,* BRE, Garston, UK.

BRE. 1984. Decay and conservation of stone masonry, *Digest 177,* BRE, Garston, UK.

BRE. 1995. Replacing wall ties, *Digest 401,* BRE, Garston, UK.

Doran, D.K. (Ed.) 1992. *Construction materials reference book,* Butterworth-Heinemann, Oxford, Ch 11. (A second edition is in preparation.)

Doran, D.K. 2009. *Site engineers manual,* 2nd edn., Whittles Publishing, Scotland, Ch 10.

Fidler, J. (Ed.). 2002. *Stone–stone building materials, construction and associated component systems: their decay and treatment,* James & James, London.

Henrya, A. and Pearce, J. 2006. *Stone conservation: principles and practice,* Donhead Publishing, Shaftesbury.

IStructE. 2008. *Manual for the design of masonry building structures to Eurocode 6,* IStructE, London.

Lynch, G. 2006. *Gauged brickwork – a technical handbook,* 2nd edn., Donhead Publishing, Shaftesbury.

Noy, E.A. 2005. *Building surveys and reports,* Blackwell Publishing, Oxford.

Smith, M.R. (Ed.) 1999. *Stone: Building stone, rock fill and armouring in construction,* Geological Society, London.

Sutherland, R.J.M. 2000. Back to the 'sixties' (1955–1975), *The Structural Engineer,* 80(6), IStructE London.

Warland, E.G. 2006. *Modern practical masonry,* Donhead Publishing, Shaftesbury.

5.2.2.2 Tudor brickwork [see text]

5.2.2.3 Terracotta and faience

Fidler, J. 1981. The conservation of architectural terracotta and faience, *Bulletin of the Association for Preservation Technology,* 15(2), Association for Preservation Technology International (APT).

5.2.2.4 Defects, repair and strengthening

BRE. 1984. *Decay and conservation of stone masonry,* BRE, Garston, UK.

BRE. 1991. *Repairing brick and block masonry,* BRE, Garston, UK.

Doran, D.K. (Ed.) 1992. *Construction materials reference book*, Butterworth-Heinemann, Oxford, Ch 49. (A second edition is in prepatation.)

Heyman, J. 1990. The maintenance of masonry. In: Somerville, G. (Ed.) T*he design life of structures*, Proceedings of the 1990 Henderson Colloquium, organised by the British Group of the IABSE, London.

5.2.2.5 Moulds, lichens and other growths

BRE. 1992. Control of lichens, moulds and similar growths, *Digest 370*, BRE, Garston, UK.

Yu, C.W. and Bull, J.W. 2006. *Durability of materials and structures: in building and civil engineering*, Whittles Publishing, Scotland.

5.2.2.6 Masonry ties

BRE. 1988. *Installing wall ties in existing construction*, Digest 329, BRE, Garston, UK.

5.2.2.7 Efflorescence

Ruddock, E.C. 1982. Cornices and pediments. In: *History of Structures*, Proceedings of the 1982 Henderson Colloquium, organised by the British Group of the IABSE, London.

Yu, C.W. and Bull, J.W. 2006. *Durability of materials and structures: in building and civil engineering*, Whittles Publishing, Scotland.

5.2.2.8 Brick chimneys

Noy, E.A. and Douglas, J. 2005. *Building surveys and reports,* 3rd edn., Blackwell Publishing, Oxford.

5.2.2.9 Brick matching and cleaning

BRE. 1983. *Cleaning of external surfaces of buildings,* Digest 280, BRE, Garston, UK.

5.2.3 Metals

5.2.3.1 Corrosion

West, J.M. 1980. *Basic corrosion and oxidation,* Ellis Horwood, Chichester.

5.2.3.2 Cast iron

Blanchard, J., Bussell, M. and Marsden, A. 1982, Appraisal of existing ferrous metal structures, *The Arup Journal*, 18(1), ARUP, London.

Bussell, M.N. 1997. *Appraisal of existing iron and steel structures,* SCI Publication No: SCI-P-183, SCI, Ascot. This publication is also applicable to wrought iron and steel.

Doran, D.K. (Ed). 1992. *Construction materials reference book,* Butterworth-Heinemann, Oxford, Ch 3. (A second edition is in preparation.)

Salmon, E.H.. 1930. *Materials and structures*, vol. 1, Longmans, London.

Stephens, J.H. (Ed.) 1976. *Structures: bridges, towers, tunnels dams,* Guinness Superlatives, Enfield.

Sutherland, R.J.M. 1982. The bending strength of cast iron. In: *History of Structures,* Proceedings of the 1982 Henderson Colloquium, organised by the British Group of the IABSE, London.

5.2.3.3 Wrought iron

Blanchard, J., Bussell, M. and Marsden, A. 1982, Appraisal of existing ferrous metal structures, *The Arup Journal*, 18(1), ARUP, London.

Doran, D.K. (Ed). 1992. *Construction materials reference book*, Butterworth-Heinemann, Oxford, Ch 4. (A second edition is in preparation.)

5.2.3.4 Steel

Note: In the period 1953–1968 the BCSA, London, produced an excellent set of guides from which practitioners will find a great deal of useful information concerning the approach to steelwork design, fabrication and erection of that era. The list includes:

BCSA. 1957. Part 4 *Examples of structural steel design Pts 1, 2 and 3.* (Superseded by Pt 13 in 1960.)

BCSA. 1952. *The collapse method of design.* (Part 5.)

BCSA. 1952. *The use of welding in steel building structures.* (Part 6.)

BCSA. 1953. *Report on experimental investigations into the behaviour of angle purlins, ties and struts.* (Part 7.)

BCSA. 1954. *Data for the use in structural steel design to conform with the requirements of BS 449: 1948.* (Part 8.)

BCSA. 1955. *Welded details for single-storey portal frames.* (Part 9.)

BCSA. 1956. *Some notes on the rigid analysis of rigid frames.* (Part 10.)

BCSA. 1957. *The collapse method of design as applied to single-bay fixed base portals.* (Part 11.)

BCSA. 1959. Part 12, *BS 449: 1959. An explanatory brochure.*

BCSA. 1960. *Examples of structural steel design to conform with the requirements of BS 449: 1959.* (Part 13.)

BSCA. 1960. *The use of welding in steel building structures.* (Part 14.)

BCSA. 1961. *Composite construction for steel framed buildings.* (Part 15.)

BCSA. 1961. *Steel frames for multi-storey buildings. Some design examples to conform with the requirements of BS 449: 1959.* (Part 16.)

BCSA. 1962. *Composite construction for steel framed buildings.* (Part 17.)

BCSA. 1962. *Notes on the use of BS 153:3A (1954) & Parts 3B & 4 (1958).* (Part 18.)

BCSA. 1963. *Deflections of portal frames.* (Part 19.)

BCSA. 1963. *Modern design of steel frames for multi-storey buildings.* (Part 20.)

BCSA. 1963. *Plastic design in steel to BS 968.* (Part 21.)

BCSA. 1963. *Examples of the design of steel girder bridges in accordance with BS 153: Parts 3A, 3B and 4.* (Part 22.)

BCSA. 1964. *The plastic design of columns.* (Part 23.)

BCSA. 1964. *Single bay single storey elastically designed portal frames.* (Part 24.)

BCSA. 1965. *Composite construction for steel framed buildings.* (Part 25.)

BCSA. 1965. *High strength friction grip bolts.* (Part 26.)

BCSA. 1965. *Design of a stanchion and truss frame.* (Part 27.)

BCSA. 1965. *Plastic design.* (Part 28.)

BCSA. 1966. *Plastic design of portal frames in steel to BS 968.* (Part 29.)

BCSA. 1967. *Safe loads and moments for stanchions to BS 449.* (Part 30.)

BCSA. 1968. *Combined bending and torsion of beams and girders.* (Part 31.)

BCSA. 1968. *The theory and practical design of bunkers.* (Part 32.)

BCSA. 1962. HS1, *The economics of structural members in high strength steel.*

BCSA. 1963. FP2, *Modern fire protection for structural steelwork.*

BCSA. 1963. M1, *Details of single bay single storey portal frame sheds.*

BCSA. 1964. M2, *Prefabricated floors for use in steel framed buildings.*

General

Bates, W. 1984. *Historical structural steelwork handbook,* BCSA, London.

BCSA *et al.,* 2007, *The prevention of corrosion on structural steelwork,* SN14 04/2007, BCSA, London.

Blanc, A. McEvoy, M. and Plank, R. (Eds.) 1993. *Architecture and construction in steel,* E & FN Spon, London.

Burdekin, F.M. 1978. Special problems in steel bridges. In: *Inspection and maintenance,* Proceedings of the 1978 Henderson Colloquium, organised by the British Group of the IABSE, London.

Doran, D.K. (Ed). 1992. *Construction materials reference book,* Butterworth-Heinemann, Oxford, Ch 5. (A second edition is in preparation.)

Historic Scotland. 2000. *Corrosion in masonry clad early 20th century steel framed buildings,* Technical Advice Note 20, Historic Scotland, Edinburgh.

Middleton, G.A.T. *c.*1900, *Modern buildings, their planning, construction and equipment,* vol. 4, Caxton Publishing, London.

Stainless steel [see text]

Weathering steel

BCSA *et al.* 2007. *The prevention of corrosion on structural steelwork,* SN14 04/2007, BCSA, London.

CORUS. 2004. *Weathering steel,* CORUS Construction and Industrial, Scunthorpe.

Cooper, M. 2007. Decongestant, *New Civil Engineer,* 22 March 2007, NCE, London. Article discusses weathering steel bridges in North Kent. Available at: http://www.nce.co.uk/decongestant/478362.article. Accessed 01 June 2009.

Welding

BCSA. 2002. *Steel bridges,* Publication No. 34/02, BCSA, London.

BCSA. 2003. *Steel building,* Publication No. 35/03, BCSA, London.

BCSA/CORUS/SCI. 2006. *Welding is a key fabrication process,* BCSA, London.

BCSA/SCI. 2003. *National structural steelwork specification for building construction,* 4th edn., No. 203/02, BCSA, London.

BCSA/SCI. 2003. *Commentary on the NSSS for building construction,* 4th edn., Publication No. 209/03, BCSA, London.

NSC. 2002. *Welding for designers,* BCSA, London.

Ogle, M.H. 1990. Design life of welded structures. In: Somerville, G. (Ed.) *The design life of structures,* Proceedings of the 1990 Henderson Colloquium, organised by the British Group of the IABSE, London.

SCI. 2002. *Guide to site welding,* SCI, Ascot.

Splash zone phenomena
CIRIA. 2005. *Managing accelerated low-water corrosion,* CIRIA, London.

Liquid metal assisted cracking (LMAC)
BCSA. 2005. *Galvanizing structural steelwork. An approach to the management of liquid metal assisted cracking,* BCSA, London.

Repair and strengthening [see text]

Steel framed and steel clad housing systems [see text]

5.2.4 Timber

Note: A great deal of general information concerning timber is to be found in Year Books issued by timber suppliers. For example, the 1969 Year Book issued by Montague L Meyer has over 250 pages of information, some of which may be of interest to practitioners investigating older buildings.

5.2.4.1 General
Bravery, A., Berry, R., Carey, J. and Cooper, D. 2003. *Recognising wood rot and insect damage in buildings,* BR453, BRE Press, Garston, UK.
BRE. 1980. *Timber decay and its control,* Technical note No 53, BRE, Garston, UK.
BRE. 1997. *Wood rot: assessing and treating decay,* GR 12, BRE, Garston, UK.
BRE. 1998. *Wood boring insect attack,* GR13, Pts 1&2, BRE, Garston, UK.
Doran, D.K. (Ed.) 1992. *Construction materials reference book,* Butterworth-Heinemann, Oxford, Ch 50. (A second edition is in preparation.)
Doran, D.K. (Ed.) 2009. *Site engineers manual,* 2nd edn., Whittles Publishing, Scotland, Ch 15.
Lyons, A. 2007. *Materials for architects and builders,* Butterworth-Heinemann, Oxford, Ch 4.
Mettem, C. Great expectations: timber repair and conservation, *The Structural Engineer,* 81(11), IStructE, London.
Wilkinson, J. and Mitchell, A. 1978. *Trees of Britain and Northern Europe,* Collins, London.

5.2.4.2 Natural defects [see text]

5.2.4.3 Building fungi and wood rot [see text]

5.2.4.4 Dry rot
BRE. 1987. *Dry rot: its recognition and control,* Digest 299, BRE, Garston, UK.

5.2.4.5 Wet rot
BRE. 1989. *Wet rots: recognition and control,* Digest 345, BRE, Garston, UK.

5.2.4.6 Insect infestation [see text]

5.2.4.7 Defective jointing (including breakdown of glued joints)

BRE. 1986. *Gluing wood successfully,* Digest 314, BRE, Garston, UK.

5.2.4.8 Metal corrosion

BRE. 1985. *Corrosion of metals by wood,* Digest 301, BRE, Garston, UK.

5.2.4.9 Repair and conservation

Doran, D.K. (Ed). 1992. *Construction materials reference book,* Butterworth-Heinemann, Oxford, Ch 50. (A second edition is in preparation.)

Mettem, C. 2003. Great expectations: timber repair and conservation, *The Structural Engineer,* 81(11), IStructE, London.

5.2.4.10 Repair and strengthening

Begg, P. 2007. Timber cantilevered staircases, *The Structural Engineer,* 85(17), IStructE, London.

British Steel. Undated. *Refurbishment in steel,* BS General Steels, Scunthorpe, UK.

Carmichael, E. 1984. *Timber engineering,* E & FN Spon, London.

Doran, D.K. (Ed). 1992. *Construction materials reference book,* Butterworth-Heinemann, Oxford, Ch 50. (A second edition is in preparation.)

IStructE. 1999. *Guide to the structural use of adhesives,* IStructE, London.

Porter, T. 2006. *Wood identification and use,* GMC Publications, Lewes, UK.

Spence R. *et al.* 2004. Whether to strengthen? Risk analysis for strengthening decision making. In: Low, A. (Ed.) *Consequences of hazards,* Proceedings of the 2004 Henderson Colloquium, organised by the British Group of the IABSE, London.

Thomas, K. 1983. Site monitoring of timber frame housing. In: Somerville, G. (Ed.) *Instrumentation of structures,* Proceedings of the 1983 Henderson Colloquium, organised by the British Group of the IABSE, London.

5.2.5 Glass

Amstock, J.S. 1997. *Handbook of glass in construction,* McGraw-Hill, Berkshire.

Button, D. *et al.* 1993. *Glass in building,* Butterworth, Oxford.

Doran, D.K. (Ed). 1992. *Construction materials reference book,* Butterworth-Heinemann, Oxford, Ch 29. (A second edition is in preparation.)

IStructE. 1999. *Structural use of glass,* IStructE, London.

Kinnear, R. 1996. *A smashing time,* Series on topical subjects, Sandberg, London.

Sobeck, W. 2005. Glass structures, *The Structural Engineer,* 83(7), IStructE, London.

Research Focus. Aug 1999. *Glass in buildings.* ICE. London.

Sedlacek, G. *et al.* 1995. Glass in structural engineering, *The Structural Engineer,* 73(2), IStructE, London.

Note: The Centre for Windows and Cladding Technology (CWCT) publish a series of Standards and Technical Reports (of which a small selection is detailed below) which will be of interest to practitioners:

CWCT. 2007. *Repairs to glass,* Report TN58, CWCT, Bath.

CWCT. 2002. *Glass in buildings: breakage – influence of nickel sulfide,* CWCT, Bath.

5.2.6 Polymers (plastics)

5.2.6.1 General

Doran, D.K. (Ed). 1992. *Construction materials reference book*, Butterworth-Heinemann, Oxford, Chs 36–47. (A second edition is in preparation.)

FOSROC. 2006. *Technical data sheet catalogue*, Fosroc International Ltd., Tamworth.

Lyons, A. 2007. *Materials for architects and builders*, 3rd edn., Butterworth-Heinnemann, Oxford.

5.2.7 Other materials of interest

5.2.7.1 Wattle and daub

Cowan, H.J. 1998. *From wattle and daub to concrete and steel: the engineering heritage of Australia's buildings*, Melbourne University Press, Carlton.

5.2.7.2 Naturally sourced materials

Straw bales

Jones, B. 2001. *Information guide to straw bale building*, Straw Bales Futures, Todmorden.

Steen, A. *et al.* 1994. *The straw bale house*, Chelsea Green Publishing, White River Junction, USA.

Bamboo

Trujillo, D. 2007. Bamboo structures in Colombia, *The Structural Engineer* 85(6), IStructE, London.

5.3 Other matters

5.3.1 Adverse environmental conditions

BRE. 1988. *Loads on roofs from snow drifting against vertical obstructions and in valleys*, Digest 332, BRE, Garston, UK.

BRE. 1994. *Wind environment around tall buildings*, Digest 390, BRE, Garston, UK.

Buller, P.S.J. 1988. *October gale 1987*, Report BR 138, BRE, Garston, UK.

Cooper, C. 2003. Living in a changing climate. In: Menzies J. (Ed.) *Climate change*, Proceedings of the 2003 Henderson Colloquium, organised by the British Group of the IABSE, London.

Fookes, P. G., Lee, E.M. and Griffiths, J.S. 2007. *Engineering Geomorphology: Theory and practice*, Whittles Publishing, Scotland. Ch 2.

Garvin, S. *et al.* 2003. Climate change implications for new buildings, In: Menzies J. (Ed.) *Climate change*, Proceedings of the 2003 Henderson Colloquium, organised by the British Group of the IABSE, London.

HMSO. 1996. *Report of the tribunal appointed to enquire into the disaster at Aberfan*, HMSO, London.

IStructE. 2000. *Seismic design and retrofit of bridges*, Reports from Seminar 22–23 June 2000, IStructE, London.

Lancaster, J.W. *et al.* 2004. *Development and flood risk*, Report C624, CIRIA, London.

Lazarus, D. 2003. *Climate change: impacts on existing buildings and historic sites.* In: Menzies J. (Ed.) *Climate change,* Proceedings of the 2003 Henderson Colloquium, organised by the British Group of the IABSE, London.

SCOSS. 2000/1. *Thirteenth report of SCOSS,* SCOSS, London.

Stansfield, K. 2001. Climate change a major structural safety issue warns SCOSS, *The Structural Engineer,* 79(13), IStructE, London.

Stansfield, K. Global warming: issues for engineers, *The Structural Engineer,* 79(14), IStructE, London.

Stern Committee. 2006. Climate change good for the economy, *The Structural Engineer,* 84(22), IStructE, London.

Webb, D. 2003. *Structural engineering in an age of climate change,* In: Menzies J. (Ed.) *Climate change,* Proceedings of the 2003 Henderson Colloquium, organised by the British Group of the IABSE, London.

5.3.2 Condensation

BRE. 1988. *Swimming pool roofs: minimising the risk of condensation using warm-deck roofing,* Digest 336, BRE, Garston, UK.

BRE. 1992. *Interstitial condensation and fabric degradation,* Digest 369, BRE, Garston, UK.

BRE. 1997. *Diagnosing the causes of dampness,* GR5, BRE, Garston, UK.

BRE. 1997. *Treating condensation in houses,* GR8, BRE, Garston, UK.

5.3.3 Dampness other than condensation

BRE. 1997. *Diagnosing the causes of dampness,* GR5, BRE, Garston, UK.

BRE. 1997. *Repairing and replacing rainwater goods,* GR9, BRE, Garston, UK.

5.3.4 Thermal insulation

BRE. 1976. *Heat losses from dwellings,* Digest 190, BRE, Garston, UK.

BRE. 1984. *Heat losses through ground floors,* Digest 145, BRE, Garston, UK.

Williams, R. and Mackechnie, C. 2008. Is this the dawning of the age of real thermal properties?, *The Building Engineer,* 83(1), Association of Building Engineers, Northampton. This article gives NPL advice on the best practice in the selection of materials for thermal insulation.

5.3.5 Sound insulation

BRE. 1988. *Sound insulation: basic principles,* Digest 337, BRE, Garston, UK.

BRE. 1988. *Insulation against external noise,* Digest 338, BRE, Garston, UK.

5.3.6 Fire

5.3.6.1 Historic background

Swailes, T. 2003. 19th century 'fireproof' buildings, their strength and robustness, *The Structural Engineer,* 81(19), IStructE, London.

5.3.6.2 Fire protection engineering

Green, M. 2002. Fire safety engineering: risk and the development of regulation. In: Thorburn, S. (Ed.) *Risk and reliability*, Proceedings of the 2002 Henderson Colloquium, organised by the British Group of the IABSE, London.

Green, M. 1998. Holistic view of fire safety engineering on the design on structures. In:

Pickett, A. (Ed.) *Structures beyond 2000*, Proceedings of the 1998 Henderson Colloquium, organised by the British Group of the IABSE, London.

Kordina, K. 1994. *Fire tests on full scale structures,* Technical University of Braunschweig. This report gives details of fire tests on single apartments (190–530 m³); flats of a dwelling house; in a tunnel and also evidence of toxic materials in tunnel fires.

5.3.6.3 Fire in concrete structures

CIRIA. 1984. *Spalling of concrete in fires,* Technical Note 118, CIRIA, London.

FIP/CEB. 1978. *FIP/CEB report on methods of assessment of the fire resistance of concrete structural members*, Cement and Concrete Association, Slough.

Lyons, A. 2007. *Materials for architects and builders,* 3rd edn., Butterworth-Heinemann, Oxford.

5.3.6.4 Fire in metal structures

Barnfield, J.R. and Porter, A.M. 1984. Historic buildings and fire: fire performance of cast-iron elements, *The Structural Engineer* 62(12), IStructE, London.

BCSA *et al.* 2006. *Steel in fire: Steel Industry Guidance Notes*, BCSA, London.

BCSA, CORUS and SCI. 2007. *Intumescent coatings*, BCSA, London.

Bond, G.V.L. *c.*1975, *Water cooled hollow columns*, Constrado SCI, Ascot.

BS 5950-8. 1990. *Code of Practice for fire resistant design*, BSI, London.

British Steel. 1985. *Checklist of intumescents available in the UK,* 5th edn., British Steel Sections, Cleveland

BSC. 1986. *The reinstatement of fire damaged steel and iron framed structures,* Swinden Laboratories, Scunthorpe.

CORUS. 2006. *Fire resistance of steel framed buildings,* CORUS, Scunthorpe.

DETR/SCI. 2000. *Fire safe design: a new approach to multi-storey steel framed buildings,* SCI, Ascot. Appendix 3 of this publication gives a brief resume of fire tests on an 8-storey steel framed building at BRE Cardington.

Eatherley, M.J. 1977. Bush Lane House, *The Structural Engineer,* 55(3), IStructE, London.

Freitag, J.K. 1903. *The fire proofing of steel buildings,* Wiley, Chapman & Hall, Toronto.

Lyons, A. 2007. *Materials for architects and builders*, 3rd edn., Butterworth-Heinemann, Oxford.

SCI. 1990. *Fire resistance design of steel structures: a handbook to BS 5950: Part 8,* SCI, Ascot.

SCI. 1991. *Investigation of Broadgate phase 8 fire, SCI,* Ascot.

Smith, D. 2006. Reliability of intumescent fire protection products, *New Steel Construction,* BCSA, London.

5.3.6.5 Fire in masonry structures

de Vekey, R. 2004. *Structural fire engineering design: materials behaviour – masonry*, BRE Digest 487, Pt. 3, BRE, Garston, UK.

Edgell, G.J. 1982. *The effect of fire on masonry and masonry structures, a review,* CERAM, Stoke-on-Trent.

Lyons, A. 2007. *Materials for architects and builders,* 3rd edn., Butterworth-Heinemann, Oxford.

5.3.6.6 Fire in timber structures

BRE. 1988. *Increasing the fire resistance of timber floors, Digest 208, BRE,* Garston, UK.

BS 5268-4. 1978. *Structural use of timber. Fire resistance of timber structures. Recommendations for calculating fire resistance of timber members,* BSI, London.

Lyons, A. 2007. *Materials for architects and builders,* 3rd edn., Butterworths, Oxford.

5.3.6.7 Effect of fire on glass

IStructE. 1999. *Structural use of glass,* IStructE, London.

5.3.6.8 Fire in plastics

Lyons, A. 2006. *Materials for architects and builders,* 3rd edn., Butterworth-Heinemann, Oxford.

5.3.7 Vibration

BRE. 2004. *The response of structures to dynamic crowd loads,* Digest 426, BRE, Garston, UK.

BRE. 1995. *Damage to structures from ground-borne vibration,* Digest 403, BRE, Garston, UK.
Note: Digest 403 is a replacement for Digest 353.

BS EN 1900: *Basis of structural design,* BSI, London

CORUS/DTI/SCI. 2007. *Design of floors for vibration: a new approach,* SCI Publication P354, SCI, Ascot.

Doran, D.K. (Ed). 1992. *Construction materials reference book,* Butterworths-Heinemann, Oxford, Ch 26. (A second edition is in preparation.)

De Silva, C.W. 2007. *Vibration: fundamentals and practice,* 2nd edn., CRC Press, London.

Devine, P.J. and Hicks, S.J. 2006. *Design guide on the vibration of floors,* 2nd edn., SCI P076, SCI, Ascot.

Jeary, A.P. 1983. *Induced vibration testing,* In: Somerville, G. (Ed.) *Instrumentation of structures,* Proceedings of the 1983 Henderson Colloquium, organised by the British Group of the IABSE, London.

5.3.8 Workmanship and site practice

BS 8000. 1989 to 1997, *Workmanship on site,* BSI, London.
Note: This BS is in several parts dealing with the following topics:
  - Excavation and filling
  - Concrete
  - Masonry
  - Waterproofing
  - Carpentry, joinery and general fixings
  - Slating and tiling of roofs and claddings
  - Glazing

- Plasterboard and dry linings
- Cement/sand floor screeds and concrete floor toppings
- Plastering and rendering
- Wall and floor tiling
- Decorative wall-coverings and painting
- Above ground drainage and sanitary appliances
- Below ground drainage
- Hot and cold water (domestic scale)
- Sealing joints in buildings and using sealants

Menzies, J.B. 1979. Workmanship. In: Bate, S.C.C. (Ed.) *Codes of practice*, Proceedings of the 1979 Henderson Colloquium, organised by the British Group of the IABSE, London.

5.3.9 General repairs

Biggs, W.D. 1987. Repair and maintenance of buildings, In: Sandberg, A. (Ed.) *Rehabilitation and renovation*, Proceedings of the 1987 Henderson Colloquium, organised by the British Group of the IABSE, London.

Ross, P. 2002. *Appraisal and repair of timber structures*, Thomas Telford, London.

Yeomans, D. 2003. *The repair of historic timber structures*, Thomas Telford, London.

5.3.10 Stability and robustness

BCSA. 2008. *Stability of temporary bracing*, SB21 01/2008, BCSA, London.

BCSA. 2004. *Code of practice for erection of low rise buildings*, Publication 36/04, BCSA, London.

BCSA. 2006. *Code of practice for erection of multi-storey buildings*, Publication 42/06, BCSA, London.

IStructE. 1989. *The achievement of structural adequacy in buildings*, IStructE, London.

5.3.11 Façade retention

CIRIA. 1994, *Structural renovation of traditional buildings*, Report R111, CIRIA, London.

CIRIA. 2003, *Retention of masonry facades*, Ref C579, CIRIA, London.

Pallett, P. 2004. Formwork, scaffolding, falsework and façade retention. In: Doran, D.K. (Ed.) *Site engineers manual*, 2nd edn., Whittles Publishing, Scotland, Ch 11.

5.3.12 Foundations

Atkinson, M.F. 2004. *Structural foundations for low-rise buildings,* 2nd edn., E & FN Spon, London.

BRE. 1993. *Damp-proof courses*, Digest 380, BRE, Garston, UK.

BRE. 1990. *Underpinning,* Digest 352, BRE, Garston, UK.

Curtin, W.G. *et al.* 2006. *Structural foundation designers manual*, Blackwell Publishing, Edinburgh.

DETR. 1997. *The Party Wall etc. Act 1996,* HMSO, London.

Jones, P.H.C. 1985. Support for St Wilfrid's Hickleston, *Concrete Magazine*, Concrete Centre, Surrey.

Pryke, J.F.S. 1987. Forms of underpinning, *Concrete magazine*, Concrete Centre, Surrey.

Serridge, C.J. 2005. Achieving sustainability in vibro stone column techniques, *Proceedings of ICE, Engineering Sustainability*, 158(4), Thomas Telford, London.

Skempton, A.W. and MacDonald D.H. 1956. The allowable settlement of structures, *Proceedings of ICE, Structures and buildings*, Structural Paper No. 50, ICE, London.

Skinner, H.D. Charles, J.A. and Tedd, P. 2005. *Brownfield sites*, Report 485, BRE, Garston, UK.

Thorburn, S. and Hutchinson, J.F. 1985. *Underpinning*, Surrey University Press, Guildford.

Tomlinson, M.J. 1987. *Foundation design and construction*, 5th edn, Longman, Harlow.

5.3.12.4 Mining subsidence

HMSO. 1951. *Mining subsidence effects on small houses*, National Building Studies Special Report No 12, HMSO, London.

NCB. 1974. *Subsidence engineers' handbook*, Mining Department, NCB, London.

Tomlinson, M.J. 2001. *Foundation design and construction*, 7th edn., Longmans, Harlow.

5.3.13 Defective basements

BS 8007: 1987. *Code of practice for design of concrete structures for retaining aqueous liquids*, BSI, London.

BS 8102: 1990. *Code of practice for protection of structures against water from the ground*, BSI, London.

Note: Includes a section giving advice on inspecting and waterproofing existing basements.

CIRIA. 1978. *Guide to the design of waterproof basements*, Westminster, London.

Gardit. 1999. *Controlling London's rising groundwater*, Thames Water Utilities, London.

IStructE. 2004. *Design and construction of deep basements including cut-and-cover structures*, IStructE, London.

IStructE. 1994. *Subsidence of low rise buildings*, IStructE, London.

SCI. 2001. *Steel intensive basements*, SCI, Ascot.

5.3.14 Liquid retaining structures

Anchor, R.D. 1992. *Design of liquid retaining structure*, 2nd Edn., Edward Arnold, London.

BS 8007: 1987. *CP for design of concrete structures for retaining aqueous liquids*, BSI, London.

Deacon, R.C. 1973. *Watertight concrete construction*, Cement and Concrete Association, Wexham Springs.

Jackson, P. 1966. Shutter ties in unlined concrete water retaining structures, *The Structural Engineer*, 44(9), IStructE, London.

Monks, W.L. 1972. *The performance of waterstops in movement joints*, Cement and Concrete Association, Wexham Springs.

5.3.15 Explosions in structures

Ellis, B.R. and Currie, D.M. 1998. Gas explosions in buildings in the UK: regulation and risk, *The Structural Engineer*, 76(19), IStructE, London.

HMSO. 1968. *Collapse of flats at Ronan Point, Canning Town*, HMSO, London.

HSE. 1985. *The Abbeystead explosion,* Report C60 2/85, HSE, Sudbury.

IStructE. 2004. *Design and construction of deep basements including cut-and-cover Structures,* IStructE, London.

Mainstone, R.J. and Butlin, R.N. 1976. *Report on an explosion at Mersey House, Bootle, Lancs,* CP 34/76, BRE, Garston, UK.

Research Focus. 1999. *Blast resistant cladding panes,* ICE, London.

SDD. 1972. *Clarkson Toll: fatal accident enquiry,* SDD, Edinburgh.

Tolloczko, J. 2006. Ways to combat site explosions, *Professional Security Magazine,* May 2006, Wolverhampton.

5.3.16　Radon gas

Daily Telegraph. 2007. *Map to detail areas at risk from cancer,* Daily Telegraph, September 2007, London.

　　Note: Presumably a somewhat different version to that supplied by the NRPB.

5.3.17　Impact damage

BS 5400-2: 1978. *Steel, concrete and composite bridges. Specification for loads,* BSI, London.

CORUS. 2005. *Bi-steel anti-attack vehicle barriers,* CORUS, Scunthorpe.

DoT/HA. 1993. Standard BD52/93, HA, London.

Molyneaux, T.C.K. *c.*1995. *Vehicle impact on masonry parapets,* University of Liverpool, Liverpool.

Tubman, J. 2004. Ship impact: flexural strain energy or good insurance. In: Low, A. (Ed.) *Consequences of Hazards,* Proceedings of the 2004 Henderson Colloquium, organised by the British Group of the IABSE, London.

5.3.18　Flat roofs

Barnes, B. 1996. *An investigation into flat roofing,* CIOB, Ascot.

BRE. 1987. *Flat roof design: thermal insulation,* Digest 324, BRE, Garston, UK.

BRE. 1986. *Flat roof design: the technical options,* Digest 312, BRE, Garston, UK.

BRE. 1992. *Flat roof design: waterproof membranes,* Digest 372, BRE, Garston, UK.

BRE. 1986. *Wind scour of gravel ballast on roofs,* Digest 311, BRE, Garston, UK.

BRE. 1998. *Flat roofs,* GR16, Parts 1&2, BRE, Garston, UK.

BRE. 1998. *Recovering flat roofs,* GR 14, BRE, Garston, UK.

Coates, D.T. 1993. *Roofs and roofing,* Whittles Publishing, Scotland.

Tarmac. 1982. *Flat roofing: a guide to good practice,* Tarmac, London.

5.3.19　Pitched roofs

BRE. 1990. *Re-covering old timber roofs,* Digest 351, BRE, Garston, UK.

Hollis, M. 2005. *Surveying buildings,* 5th edn., RICS Books, London.

5.3.20　Windows and Doors

BRE. 1997. *Draughty windows, condensation in sealed units, operating problems,* GR10, BRE, Garston, UK.

Research Focus. 2001. *Developing a UK domestic window energy rating system,* RF No.45.

Totton, M. and Hirst, E. (Eds.) 2007. *Windows: history, repair and conservation*, Donhead Publishing, Shaftesbury.

5.3.21 Slip resistance of floors

HSE. 2007. *Assessing the slip resistance of flooring*, Information Sheet, Health and Safety Executive, Bootle.

5.3.22 Tenements

Mcleod, I.A. 1989. Scottish tenements: a case study in structural repair, *The Structural Engineer*, 67(2), IStructE, London.
  Note: This paper gives advice on structural state of buildings; advice on alterations and inspection of structural matters during reconstruction.

5.3.23 Heritage and Ecclesiastical buildings (including burial rights)

Heyman, J. (Ed.) 1995. *The stone skeleton*, Cambridge University Press, Cambridge.
Heyman, J. 1996. *Arches vaults and buttresses*, Ashgate Variorium, Aldershot.
Jones, L.E. 1965. *Old English churches*, Frederick Warne, London.
Lilley, D.M. and March, A.V. *c.*2004, *Problems in ecclesiastical buildings resulting from medieval burial practices*, Newcastle University, Newcastle. (Contact d.m.lilley@ncl.ac.uk)
Tomsett, H.N. 2010. *Handbook of non-destructive testing in construction*, Whittles Publishing, Scotland. (In preparation.)

5.3.24 Bridges

Bakht, B. 1988. *Actual versus assumed behaviour of girder bridges*, 5th ASCE Speciality Conference on Probalistic methods in Civil Engineering May 1988, ASCE.
Bakht, B. and Jaeger, L. 1988. Bearing restraint in slab-on-girder bridges, *ASCE Journal of Structural Engineering*, 114(12), ASCE.
Bakht, B. and Jaeger, L. 1990. Bridge evaluation for multipresence of vehicles, *ASCE Journal of Structural Engineering*, 116(3), ASCE.
Bakht, B. and Mufti, A. 1991. *Evaluation of a deteriorated concrete bridge by testing*, Seminar paper concerning research undertaken for the Ministry of Transportation of Ontario.
Bakht, B. and Mufti, A. 1991. *Proof load test on a short- span bridge*, Seminar paper concerning research undertaken for the Ministry of Transportation of Ontario.
Brown, D.J. 1996. *Three thousand years of defying nature*, Mitchell Beazley Octopus Publishing, London.
Burdiken, F.M. 1978. Special problems in steel bridges. In: *Inspection and maintenance*, Proceedings of the 1978 Henderson Colloquium, organised by the British Group of the IABSE, London.
Bulson, P.S. 1981. Future Military Engineering Technology, *The Royal Engineers Journal* 95(1).
Bulson P.S. 1985. Military bridges. In: Wood, D. (Ed.) *Movable structures*, Proceedings of the 1985 Henderson Colloquium, organised by the British Group of the IABSE, London.

CIRIA. 2006. *Masonry arch bridges: condition, appraisal and remedial treatment.* Report C656, CIRIA, London.

DoT/HA. 2001. *Assessment of masonry arch bridges. Modified MEXE method,* vol. 3,Section 4, Part 4. BA 16/97 Amd No:2. HA. London.

Foss, C.F. and Gander, T.J. (Eds.) 1986. *Jane's military vehicles and ground support equipment,* Jane's Information Group, Coulsdon.

Hall, M. 2005. *Earth building: Methods and materials, repair and conservation,* Taylor & Francis, London.

Heyman, J. 1982. *The masonry arch,* Ellis Horwood, Chichester.

HMSO. 1971. *Inquiry into the basis of design and method of erection of steel box girder bridges,* Interim Report, HMSO, London.
Note: This report was commissioned as a result of current problems on the Milford Haven bridge and others. Recommendations contained in this report led in time to the formulation and promulgation of the so-called Merrison rules for the design of steel box girder bridges.

HMSO. 1984. *The assessment of highway bridges and structures,* BD21/84, HMSO, London (now superseded).

HMSO. 1988. *Loads for highway bridges,* BD37/88, HMSO, London (partly superseded).

Hogland, T. 2006. Aluminium in bridge decks and in a New Military Bridge in Sweden, *Structural Engineering International,* 16(4), IABSE, Zurich.

ICE. 1995. *The use of plate bonding: report of half day meeting,* ICE, London.

ICE. 2007. *Thomas Telford. 250 Years of Inspiration,* ICE, London.
Note: This publication gives details of Telford's bridge history including his Cast Iron, Masonry and Suspension bridges.

IStructE. 2001. *Guide to inspection of underwater structures,* IStructE, London.

Jaeger, L. and Bakht, B. 1987. *Multiple presence reduction factors for bridges,* ASCE Structures Congress, Orlando.

Ko, R. 2004. Vehicle collision loading criteria for bridge piers. In: Low, A. (Ed.) *Consequences of hazards,* Proceedings of the 2004 Henderson Colloquium, organised by the British Group of the IABSE, London.
Note: Paper concerning research carried out on behalf of the Ministry of transportation of Ontario, Canada.

Kumar, P. and Bhandari, N.M. 2006. Mechanism based assessment of masonry arch bridges, *Structural Engineering International,* 3/2006, IABSE, Zurich.

Leonhardt, F. 1982. *Brucken Bridges,* Architectural Press, London.

Long, A. 2007. *Sustainable bridges through innovative advances,* ICE & TRF Fellows Lecture 2 May 2007, ICE, London.

Melbourne, C. c.2000, *The behaviour of masonry arch bridges,* Bolton University. Bolton, Lancashire.

Military Engineering. 1956. *Bailey bridge–normal use,* HMSO, London.

Ministry of Transportation. 1988. *Ultimate load test of a slab-on-girder bridge,* Report SRR-888-03, Ministry of Tranport, Ontario.
Note: This is a report on a full scale load test on the 40-year-old Stoney Creek Bridge, City of London, Ontario before its replacement with a wider bridge.

Ministry of Transportation. 1988. *Testing of an old short span-on-girder bridge,* Report SRR-88-01, Ministry of Transport, Ontario.
> **Note:** This is a report of load test on the deteriorating 1953 16.3 m span, Belle River Bridge in the County of Essex, Canada.

Ministry of Transportation. 1988. *Observed behaviour of a new medium span slab-on-girder bridge,* Report SRR-88-02, Ministry of Transport, Ontario.
> **Note:** This is a report on a load test on the 45.72 m span North Muskoka River Bridge which was about 13 years old at the time of test.

Ministry of Transportation. 1989. *Review of dynamic testing of highway bridges,* Report SRR-89-01, Ministry of Transport, Ontario.

Mufti, A. *et al. c.*1991. *Experimental investigation of FRC (fibre reinforced concrete) deck slabs without internal steel reinforcement,* Seminar paper based on experimental work carried out under the auspices of the NSERC of Canada and the Ministry of Transportation of Ontario.

Royal Commission. 1971. *Failure of Westgate bridge. Report: Victoria, Australia,* CH Rixon, Melbourne.
> **Note:** This report describes a classic example of how things can go wrong when site staff modify construction procedures without adequate briefing from the designer. In the opinion of the Editor this report should be essential reading for any who aspire to join the ranks of construction professionals.

Redfern, B. 2006. Cast iron conundrum, *New Civil Engineer,* 5 October 2006, *Emap Construct Ltd.,* London.

Rigden, S.R. 1996. Long term performance of concrete bridges, *Construction Repair 10(4).*
> **Note***: The Construction Repair* magazine as been subsumed into *Concrete Engineering International* and enquiries should be addressed to The Concrete Society, Surrey.

Sowden, A.M. 1978. The organisation of the inspection of bridges in British Railways. In: *Inspection and Maintenance,* Proceedings of the 1978 Henderson Colloquium, organised by the British Group of the IABSE, London.

Sumon, S.K. 2005. Innovative retrofitted reinforcement techniques for masonry arch bridges, *Proceedings of ICE, Bridge Engineering,* 158(3), Thomas Telford, London.

Tajalli, S.M.A. and Rigden, S.R. 2000. Partially and non-destructive testing of 40 concrete bridges, *Proceedings of ICE, Structures and Buildings,* 140(1), Thomas Telford, London.

Thompson, D. *et al.* 1993. Continuous arches. In: Pritchard, B. (Ed.) *Towards joint-free bridges,* Proceedings of the 1993 Henderson Colloquium, organised by the British Group of the IABSE, London.

Tomor, A.K. and Melbourne, C. 2007. Condition monitoring of masonry arch bridges using acoustic emission techniques, *Structural Engineering International,* 2(17), IABSE, Zurich.

Williams, M. 2006. Wind Shielding for bridge decks, *The Structural Engineer,* 84(10), IStructE, London.

5.3.25 Tunnels

Blake, L.S. 1989. *Civil engineer's reference book,* 4th edn. Butterworth-Heinemann, Oxford, Ch 32.

Faure, R.M. and Karray, M. 2007. Investigation of the concrete tunnel lining after the Mont Blanc fire, *Structural Engineering International*, 2/2007, IABSE, Zurich.

HSE. 1996. *Safety of New Austrian tunnelling method (NATM) tunnels. A review of sprayed concrete lined tunnels with particular reference to London clay*, HSE, Bootle.

IStructE. 2004. *Design and construction of deep basements, Appendix E.3 Recent developments in using the Observational Method*, IStructE, London.

Stanway, L.C. 2000. *The story of the carriage of the mails on London's underground railways*, AEPS, Basildon.

Stanway, L.C. 2002. *Mails under London*, AEPS, Basildon.

5.3.26 Cladding

Anderson, J.M. and Gill, J.R. 1988. *Rain-screen cladding: a guide to design principles and practice*, Butterworth, Oxford.

BCSA/CORUS/SCI. 2006. *Achieving air-tightness with metal cladding systems*, SN06 06/2006, BCSA, London.

BRE. 1978. *Wall cladding defects and their diagnosis*, Digest 217, BRE, Garston, UK.

BSI, CP 298: 1972. *Natural stone cladding (non-load-bearing)*, BSI, London.

BS 8298: 1994. *Code of practice for design and installation of natural stone cladding and lining*, BSI, London.

BS 5427-1: 1996. *Code of practice for the use of profiled sheet for roof and wall cladding on buildings*, BSI, London.

BS 8297: 2000. *Code of practice for design and installation of non-load-bearing precast concrete cladding*, BSI, London.

BS EN 12326: 2000, *Slate and stone products for discontinuous roofing and cladding. Methods of test*, BSI, London.

CS. 1977. *Guide to pre-cast concrete cladding*, Technical Report No. 14, Concrete Society, Surrey.

CWCT. 1999. *Testing of fixings for thin stone cladding*, CWCT, Bath.

Doran, D.K. (Ed). 1992. *Construction materials reference book*, Butterworth-Heinemann, Oxford. (A second edition is in preparation.)

Harrison, H.W., Hunt, J.H. and Thomson, J. 1986. *Overcladding exterior walls of large panel system dwelling*, BRE, Garston, UK.

IStructE. 1989. *Guidance note on the security of the outer leaf of large concrete panels of sandwich construction*, IStructE, London.

IStructE. 1995. *Aspects of cladding*, IStructE, London.

Ledbetter, S. 1997. Serviceability in building cladding. In: Nethercot, D. (Ed.) *Structures for serviceability*, Proceedings of the 1997 Henderson Colloquium, organised by the British Group of the IABSE, London.

Research Focus. 2000. *Design guidelines for overcladding systems to maintain durability in the building fabric*, RF No. 43, ICE, London.

Ryan, P.A. and Wolstenholme, P. 1995. *Industrial cladding, assessment and repair, Construction Repair Magazine*.

Note: *The Construction Repair* magazine as been subsumed into *Concrete Engineering International* and enquiries should be addressed to The Concrete Society, Surrey.

Ryan, P.A. *et al.* 1994. *Durability of cladding – a state of the art report*, Thomas Telford, London.

Taylor, H.P.J. 1998. Cladding. In: Pickett A. (Ed.) *Structures beyond 2000,* Proceedings of the 1998 Henderson Colloquium, organised by the British Group IABSE, London.

Taywood *et al..* 1999. *Cladding Buildability* (in CD format), DETR, Taywood, Leighton Buzzard.

Taywood/DETR. 1999. *The prevention and repair of corrosion in masonry clad steel framed buildings* (in CD format), DETR Ref MT2C, Taywood, Leighton Buzzard.

Wilson, M. and Harrison, P. 1993. *Appraisal and repair of claddings and fixings,* Thomas Telford, London.

5.3.27 Asbestos

Doran, D.K. (Ed). 1992. *Construction materials reference book,* Butterworth-Heinemann, Oxford, Ch 9. (A second edition is in preparation.)

HSE. 2002. *A short guide to managing asbestos in premises,* HSE, Bootle.

HSE. 1999. *Working with asbestos in buildings,* HSE, Bootle.

5.3.28 Japanese knotweed

Jay, M. 2006. *Japanese knotweed in the construction industry,* TBE, Northampton.

HMSO. 1981. *Wildlife and Countryside Act,* HMSO, London.

HMSO. 1959. *Weeds Act,* HMSO, London.

5.3.29 Services installations

CIBSE. 2008. Publications matrix, CIBSE, www.cibse.org. Accessed 01 May 2009.

Note: This is a useful one-stop guide to services provided by the CIBSE.

5.3.30 Underground services

BRE. 1984. *Access to underground drainage systems,* Digest 292, BRE, Garston, UK.

Downey, D. 2006. Trenchless technology: a modern solution for clean flowing cities. *Proceedings of ICE, Civil Engineering,* 160(2), Thomas Telford, London.

Richardson, L. 2009. Drains, Sewers and Services. In: Doran, D.K. (Ed.) *Site Engineers Manual,* 2nd edn., Whittles Publishing, Scotland.

6 Legal Restraints

6.1 Planning

Existing Town & Country Planning legislation has been consolidated into three acts; *The Town & Country Planning Act 1990*; *The Planning (Listed Buildings and Conservation Areas) Act 1999*; and *The Planning (Hazardous Substance) Act 1990*.

In addition, *The Ancient Monuments and Archaeological Areas Act 1979* might apply to some situations.

6.1.1 Development control

The major form of development control in the United Kingdom is carried out through the planning system. Planning policy is set by central government (or devolved government) and implemented locally. For most developments and changes of use of buildings and land use planning permission will be required.

Application for planning permission has to be made to the local authority and they will judge the application against locally set guidelines.

Some areas and buildings might have historic or architectural significance. Buildings can be listed and areas can be granted Conservation Area Status. The control of development in these cases is also carried out through the local authority planning system but requires Listed Building Consent and/or Conservation Area Consent in addition to normal planning permission. Buildings and other structures designated as Ancient Monuments may also require special consideration. These are covered by The Ancient monuments and Archaeological Areas Act 1979; further advice may be obtained from English Heritage (EH).

For many frequently made applications there may be local 'permitted development rights' covering work such as extending a house. Within the locally set guidelines these works are permitted under the planning rules without having to make an application. In some areas of historic or architectural interest permitted development rights can be severely restricted and detailed control over paint colours and types of fencing might be imposed.

Consultation with the planning department at the local authority should establish whether formal applications are required for the works proposed. If there is uncertainty about the current status of a building an application can be made for a 'lawful development certificate'. This will formally set out the view of the local authority.

6.1.2 Planning application

In general terms, making a planning application is very straightforward. Prepare drawings and other information about the proposed works, fill in the application form, pay the planning application fee, and submit the application to the local authority and wait for approval. The local authority has to consult with other parties such as the Highways Department.

6.1.3 The legal position

In most cases it is not illegal to carry out a development before applying for or receiving planning permission. It is however a very unwise route to take as the local authority could decide not to grant permission and require a return to the original condition. It is an offence to demolish, alter or extend a listed building without listed building consent and the penalty can be a fine of unlimited amount or up to twelve months' imprisonment, or both. Planning permission is not sufficient to allow demolition of a Listed Building.

With a few exceptions the protection granted to Listed Buildings is extended to buildings in a Conservation Area.

6.1.4 Check list

Responsibilities:

- Establish who has responsibility for permissions under the planning process.
- At different stages responsibility might change from the client to consultants to contractors.
- Responsibility for complying with any conditions attached to approvals might also change.
- Identify critical path items. Some permissions and conditions attached to them might influence, or even determine a programme of work.
- Establish how the progress will be monitored.

Planning permission

- Check if permission is required.
- Check if an application has been made or granted.
- Find out the conditions that apply to any permission.
- Find out at what stage conditions have to be complied with, some might be before the development starts.

Are any of the structures Listed Buildings?

- Check if Listed Building Consent is required.
- Check any conditions that apply to the consent.

Is there a Conservation Area?

- Check if Conservation area consent is required. Local authorities have a duty to designate these areas if they are construed to be of special architectural interest. As a consequence only certain styles and patterns of development are permitted, thus preserving the general ambience of the area.
- Check any conditions that apply to the consent.

Additional factors to be considered are any restrictions that might apply to the development of a World Heritage Site. One such in the UK is the Cornwall and West Devon mining landscape. By 2006 23 UK sites had been granted this status.

Trees and other features

- Do any of the trees have Tree Preservation Orders (TPOs) on them? These are orders, which may be granted by local authorities, to limit destruction and preserve appearance of certain trees. In general only tree surgeons or others registered with the Arboricultural Association are permitted to work on trees covered by a TPO
- The planning condition may include protection of other trees and features on the site.

Other authorities

Building Regulations approval will be required and may affect some aspects of planning approval. Carrying out work without Building Regulations approval is illegal.

Highways and other road authorities might have an effect on some aspects of the works. Consultation would normally be carried out through the planning application process.

The local Fire Brigade might have an influence on a scheme and would normally be consulted as part to the Building Regulations application process. Any appeal against the above Regulations requires an approach to The Secretary of State (Planning Inspectorate).

Environmental Health Departments could have an influence on a project including aspects of food safety, controlling pollution, contamination, noise and safe working environments.

6.2 Listed Buildings

At the outset of any project, establish if a building is a Listed Building. If it is, it has protection against unauthorised demolition, alteration and extension. Carrying out unauthorised work on a listed building is a criminal offence punishable by a fine or a prison sentence. The building might also have to be restored to its previous state.

Buildings are listed because they are of special architectural or historical interest or both. This can apply to an individual building or a group of buildings. To make changes to a listed building that would affect its character, inside or out, it must first receive listed building consent. An application for consent is made through the planning department at the relevant local authority.

There are three grades of listed buildings depending on their importance. In England and Wales Grade I buildings are those of exceptional interest. Grade II* (grade two star) are particularly important buildings of more than special interest and Grade II of special interest. In Scotland and Northern Ireland Grades I, II* and II are replaced by Grades A, B and C.

6.2.1 Applying for Listed Building Consent

The local authority will deal with most applications but the most important cases are referred to other authorities. The authority in England is English Heritage. In Wales, the statutory body is Cadw. In Scotland it is Historic Scotland. The Ulster Architectural Society looks after Northern Ireland's listed buildings. Important cases can be referred to higher levels of government.

In England local authorities have to notify English Heritage when they first receive applications affecting buildings of outstanding national interest, normally those listed Grade I and II*. Local authorities may refuse any listed building consent applications but they may not grant consent for any works to a Grade I or II* building, or substantial demolition of a Grade II building, without first referring the case to the Secretary of State to consider whether to 'call in' the applications. In these cases the advice of English Heritage will be asked for and if the application is called in it is usual to hold a public inquiry. Planning permission might be needed for some projects in addition to listed building consent.

In Greater London, English Heritage has the power to direct the Boroughs' decisions on all listed building consent applications. In a number of Boroughs power has been delegated for them to deal with proposals for minor alterations and extensions to Grade II listed buildings.

Repairs to a Listed Building carried out to match exactly the original work may not need consent, but always consult the local authority for advice. The judgement on whether a repair is suitable is not always straightforward.

Examples of works which are likely to need consent include changing windows and doors, painting over brickwork or removing external surfaces, putting in dormer windows or rooflights, putting up aerials, satellite dishes and burglar alarms, changing roofing materials, moving or removing internal walls, making new doorways, and removing or altering fireplaces, panelling or staircases.

6.2.2 The listing of buildings

English Heritage now administers the listing of buildings in England. They make recommendations to the Secretary of State about whether to add buildings to the statutory list. Anyone can apply to English Heritage for a building to be listed. An appeal can also be made against a decision to list a building.

A listed building can be removed from the list if it no longer meets the required criteria. Only architectural or historic interest will be taken into consideration. If an application has been made for listed building consent, or an appeal made against refusal of consent, or if action by a local planning authority is in hand de-listing would not be considered.

Building Preservation Notices (BPN) can be served by planning authorities and National Park authorities on the owner of a building that is not listed. This would protect the building for a period of six months as if it were in fact listed. This allows time for a proper consideration of the architectural and historic merits.

A Certificate of Immunity (COI) can be issued to prevent a building being listed for a period of five years. This prevents the planning authority from serving a BPN for that period. If planning permission has been applied for or granted a COI may be requested from the secretary of state. The certificate gives greater certainty when works are proposed to a building that may be eligible for listing. If a COI is not granted, then a building will normally be added to the statutory list.

6.2.3 Other legislation

There is also a Schedule of Ancient Monuments. In the rare cases that a building is both scheduled and listed, ancient monuments legislation takes precedence.

Special rules apply to churches that are listed buildings. Certain denominations have their own system of control and approval over work to churches still in ecclesiastical use.

Conservation areas can be set up if they are considered 'areas of special architectural or historic interest, the character or appearance of which it is desirable to preserve or enhance'. Consult the local authority to find out what rules apply to buildings within a conservation area. Within the area, buildings have similar protection from demolition as that given to listed buildings.

Value added tax does not apply to the cost of alterations to listed buildings, although it does apply to repairs and ordinary maintenance. More information is available from VAT offices in the leaflet VAT: Protected buildings (708/1/90). Grade I and II* buildings may be eligible for English Heritage grants for urgent major repairs. Grants for a Grade II listed building are unlikely to be given. Other sources of funding might be available including those from special interest groups.

6.3 Building Legislation

6.3.1 Local Acts

A number of Local Acts (Parliamentary Acts promoted by a Local Authority) still exist. These Acts can affect the design and/or construction of a building and the Building Control Officer should be asked for details of those in force in his area. It should be noted that Approved Inspectors (Private Building Control Organisations registered with CIC) cannot deal with Local Acts. The best known Local Acts are the series of *London Building Acts 1930-1939*.

Until 1986 Inner London Building Control was administered differently to the remainder of England in accordance with the *London Building Acts and Bylaws*. Control was exercised by District Surveyors with discretionary powers. *The Building (London) Act 1985* changed the London system to bring it into line with National Regulations and with the demise of the GLC in 1986, control transferred to the inner London Boroughs. However significant sections of the *London Building Acts* remain in force in Inner London, particularly *The London Building Acts (Amendment) Act 1939 Section 20 Fire Safety in Buildings.*

This Act applies where: (a) a building is to be erected with a storey or part of a storey at a greater height than (i) 30 m or (ii) 25 m if the area of the building exceeds 930 m², and (b) a building of the warehouse class, or a building or part of a building used for the purposes of trade or manufacture which exceeds 7100 m³ in extent unless it is divided by division walls in such a manner that no division of the building is of a cubical extent exceeding 7100 m³. The principle areas it covers are fire alarms, automatic fire detection system, the installation of fire extinguishing apparatus and appliances, smoke ventilation of a building and access to the building by the fire brigade personnel.

Measurement of the height of any storey or part of a storey must be taken at the centre of that face of the building where the height is the greatest from the level of the pavement immediately in front of that face or where there is no such pavement from the level of the ground before excavation to the level of the highest part of the interior of the top storey.

Section 30 – Special and Temporary Buildings and Structures applies to the erection or retention of certain temporary buildings that are not covered by the Building Regulations 2000, i.e. temporary stands or similar structures.

Part VII – Dangerous and Neglected Structures: If the building or structure becomes dangerous to the public, the local authority has powers to deal with and take emergency measures for the situation. In such cases, the District Surveyor will issue a Dangerous Structure Notice. This provision applies to the whole of Greater London except Barnet.

In areas outside London, the Local Authority operating through the Building Control Officer has similar powers under the Building Regulations.

6.3.2 The Party Wall etc. Act 1996

The opening statement of this 15-page legal document is:

> An Act to make provision in respect of party walls, and excavation and construction in proximity to certain buildings or structures; and for connected purposes. [18th July 1996]

Its purpose is to protect the rights of adjoining property owners from the wrongful acts of their neighbours. The Act deals with topics including:

- party walls and party structures (a party structure might for example be a floor between two apartments)
- excavations within 3–6 m of adjacent property (see Figs. 6.1 and 6.2)
- surcharging of adjacent foundations
- prior notice to adjoining owner
- rights of entry into adjoining property
- resolution of disputes

6.3.3 Health and Safety

Health and Safety requirements are enshrined in law. Non-compliance may lead to prosecution and make it difficult to defend civil claims. It is therefore essential that all site personnel become acquainted with and comply with the basic legal requirements.

Practitioners should note that under the *Health and Safety at Work Act* (HSWA) it is now mandatory for all visitors on site to be in possession of a visitor's pass (see Fig. 6.4).

Figure 6.3 An example of poor safety arrangements for the worker (Courtesy HSE).

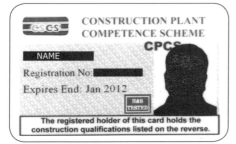

Figure 6.4 Construction site visitor's pass (*left*: front view; *right*: reverse view).

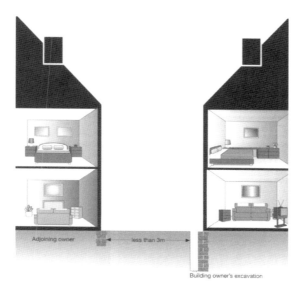

Figure 6.1 Party Wall etc. Act: 3 m rule (Courtesy ODPM).

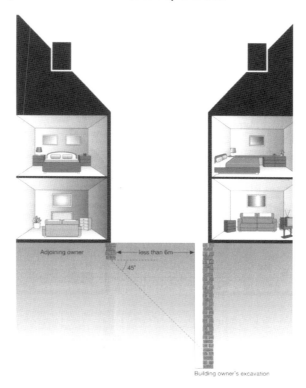

Figure 6.2 Party Wall etc. Act: 6 m rule (Courtesy ODPM).

To many, health and safety requirements are somewhat tedious. However, it should not be forgotten that on average 70 site personnel die each year from asbestos related diseases. Recent legislation moving towards acceptance of the concept of Corporate Manslaughter is worrying to senior executives of contracting and consultancy organisations.

It should be remember, however, that Health and Safety is not a new concept, for the Bible states:

> *When you build a new house, put a parapet around the roof. Otherwise, if someone falls off, you will bring bloodguilt upon your house.*
> — Deuteronomy, ch 22, v 8.

It is also worth noting that HSE have recently reported that more than half the fatalities that occur in construction are related to refurbishment work. Many of these rules are derived from *The Health and Safety at Work etc. Act 1974.*

At the time of writing there are over 30 sets of regulations which might apply to construction. These include:

- Health and Safety at Work etc Act (HSWA)
- Construction (Head Protection) Regulations (C(HP)R)
- Construction (Health, Safety and Welfare) Regulations (C(HSW)R)
- Lifting Operations and Lifting Equipment Regulations (LOLER)
- Construction (Design and Management) Regulations 1994 (CDM) (for more details see elsewhere in this chapter)
- Management of Health and Safety at Work Regulations (MHSWR)
- Manual Handling Operations Regulations (MHOR)
- Personal Protective Equipment at Work Regulations (PPE)
- Control of Substances Hazardous to Health Regulations (COSH)
- The Fire Precautions Act (FPA)
- Abrasive Wheels Regulations (AWR)
- Health and Safety (First Aid Regulations) (HS(FAR))
- Reporting of Injuries, Diseases and Dangerous Occurrences Regulations (RIDDOR)
- Control of Asbestos at Work Regulations (CAWR)
- Control of Lead at Work Regulations (CLWR)
- Noise at Work Regulations (NWR)
- Electricity at Work Regulations (EWR)
- Provision and Use of Work Equipment Regulations (PUWER)
- The Ionising Radiations Regulations (IRR)
- Diving at Work Regulations (DWR)
- Health and Safety (Safety Signs and Signals) Regulations (HS(SSS)R)

This topic is covered in more detail in Chapter 19 of *Site Engineers Manual – 2nd edition* (Doran *et al.*, 2009).

6.3.4 Disability Discrimination Act 2005

This 69-page document contains 20 Clauses and 2 Schedules with provisions to amend the Disability Discrimination Act 1995 (DDA) and builds on amendments already made to that Act by other legislation.

The provisions extend generally to Britain and are grouped under headings relating to public authorities, transport and other matters. Provisions include the creation of a new duty on public bodies to have due regard to the need to eliminate unlawful discrimination and harassment of disabled people. Explanatory notes are available together with copies of the Act from The Stationery Office.

6.3.5 The Regulatory Reform (Fire safety) Order 2005 (RRO 2005)

This Act came into force on 1 October 2006 and repeals or amends over 70 pieces of existing legislation. The list includes some the most significant changes:

- Fire certificates will no longer be valid.
- The new focus will be on risk assessment.
- Compliance with RRO 2005 is the duty of the responsible person – possibly the owner, employer, occupier or lessee.
- Employees are subject to general duties, including a duty to co-operate and to alert the employer to certain risks.
- In multi-occupied buildings, the owners and occupiers of other parts of the building are required to co-operate with the responsible person in making arrangements for the maintenance of facilities, equipment and devices for fire safety.
- Duties in respect of fire safety are owed not only to employees but also to relevant persons which includes anyone lawfully on the premises or in the vicinity of the premises and at risk from fire at the premises.
- If premises are subject to separate licensing control (for example theatres and sports grounds) fire safety requirements specified in the licence must be consistent with those contained in RRO 2005.
- The department for Communities and Local Government (DCLG) is required under RRO 2005 to issue guidance to assist responsible persons in carrying out their duties under RRO 2005.
- Enforcement of the provisions of RRO 2005 remains the responsibility of the local fire authority for most types of premises. Sanctions for failure to comply with RRO 2005 include fines and imprisonment.

6.3.6 Warranties

See Chapter 4 for Collateral Warranties.

6.3.7 Building Acts and Regulations in England

These are mandatory regulations with which a developer must comply. Complete sets of Building Regulations are available from the Stationery Office and a Government explanatory booklet (*Building Regulations*) is available, free of charge. In England the Building Regulations are enforced by the local Building Control Officer (sometimes known as the District Surveyor). A more recent innovation is the establishment of the Independent Inspector, a specialist consultant set up to provide competition to the local authority Building Control Officer, who has the authority to check and approve matters of Building Regulations. However an Independent Inspector does not have the authority to stop work in progress if non-compliance is suspected. Such action could only be arranged in conjunction with the local authority. As a matter of interest, in Jersey, a new system – The Structural Engineers Registration Ltd (SER Ltd) – has been established as an alternative Building Control system. For further information on this and other Scottish building control matters, the reader is directed to the appropriate website(s).

The following is a list of the various non-mandatory Approved Documents (ADs):

- Approved Document A – Structure
- Approved Document B – Fire Safety
- Approved Document C – Site Preparation and Resistance to Moisture
- Approved Document C – Toxic Substances
- Approved Document E – Resistance to Passage of Sound
- Amendments to Approved Document E – Resistance to Passage of Sound
- Approved Document F – Ventilation
- Approved Document G – Hygiene
- Approved Document H – Drainage and Waste Disposal
- Approved Document J – Guidance and supplementary information on UK implementation of European Standards for Chimneys and Flues
- Approved Document K Protection from falling, Collision Impact
- Approved Document L1 – Conservation of Fuel and Power in Dwellings
- Approved Document L2 – Conservation of Fuel and Power Buildings Other than Dwellings
- Approved Document M – Access to and Use of Buildings
- Approved Document N – Glazing – Safety in Relation to Impact Opening and Cleaning
- Approved Document P – Electrical Safety
- Approved Document to Support Reg. 7 – Materials and Workmanship
- Building Regulations and Fire Safety: Procedural Guidance
- Statutory Instrument 2001/3335
- DTLR Circular 3/2001
- Limiting Thermal Bridging and Air leakage, Robust Construction Details for Dwellings and Similar Buildings

- Thermal Insulation Avoiding Risks, a Good Practice Guide Supporting Building Regulation Requirements 2002, 3rd edition.
- Span Tables for Solid Timber Members for Dwellings
- Private Approved Document – Basement for Dwellings

It should be noted that, in the recent amendments to Part A of these regulations, a new category of buildings has been introduced under the classification 'Class 3'. This includes hospitals over three storeys; most other buildings over 15 storeys; public buildings over 5000 sq metres and stadiums accommodating more than 5000 spectators. These buildings or structures will require a systematic risk assessment of foreseeable hazards.

6.3.8 Landfill (England and Wales) Regulations 2002 (as amended)

This 42-page document deals with, amongst other things, definition of landfill, dredgings, planning permission, landfills of different classes in close proximity, permit conditions, implementation, liquid waste, waste treatment, acceptance procedures, costs, aspects of closure, training, and enforcement. Documentation is available through Defra.

6.3.9 Asbestos

For more information see Chapter 5, Section 5.3.27.

6.3.10 Reservoirs Act 1975

Safety legislation for reservoirs in the United Kingdom was first introduced in 1930 after several reservoir disasters had resulted in loss of life. This early Act was superseded by the Reservoirs Act 1975, which now provides the legal framework to ensure the safety of large raised reservoirs and applies to reservoirs that hold at least 25,000 cubic metres of water above natural ground level.

Under the Reservoirs Act 1975 reservoir owners (Undertakers) have ultimate responsibility for their reservoirs. They must appoint a specialist civil engineer (who is qualified and experienced in reservoir safety) to continuously supervise the reservoir (Supervising Engineer) and to carry out periodic inspections (Inspecting Engineer). A Panel Engineer must also be appointed to design and construct a new reservoir or repair or make changes to an existing reservoir (Construction Engineer).

A periodic inspection by an Inspecting Engineer is required every ten years, or more frequently if necessary. As a result of his inspection, he will specify a safe operating regime and he may recommend works required 'in the interests of safety'.

A Supervising Engineer is required to supervise the operation and maintenance of the reservoir and produce an annual statement. He can recommend that a periodic inspection is carried out.

For reservoirs below the threshold of 25,000 cubic metres regulation is managed by the Health and Safety Executive.

The Institution of Civil Engineers retains a list of civil engineers qualified and registered to be involved in this type of work.

6.3.11 Construction (Design and Management) Regulations 1994 (CDM), revised April 2007

In response to EU Directive 92/57 EEC the CDM Regulations were introduced in 1995. The aim was to reduce the unacceptable level of accidents on construction sites. Clients and designers became part of the safety process and were given statutory duties in the regulations.

A significant revision of the regulations in 2007 redefined some roles, responsibilities and procedures. Co-operation, resourcing and competence are an essential part of the regulations. The regulations now apply to all construction projects and, as before, this broad definition includes some maintenance work and dismantling of plant.

For most construction work the client must appoint a CDM co-ordinator (previously a planning supervisor) at the early stages of a project. The role of the CDM co-ordinator includes the co-ordination of the health and safety aspects of design work and notifying the Health and Safety Executive of the project. The CDM co-ordinator has to produce pre-construction health and safety information so designers and contractors are aware of hazards that might be faced on a project.

Before construction work starts the client must appoint a Principal Contractor to take overall responsibility for health and safety and for the production of a health and safety plan for the construction phase of the work. At the end of construction work, information on the project must be provided to the client including a health and safety file produced by the CDM co-ordinator.

Small projects, lasting less than 30 days or 500 person days, and work for domestic clients are exempt from some parts of the regulations.

For further information see the References and regulations.

6.3.12 The Construction Products Directive (CPD)

This is a European Directive that seeks to remove barriers to trade. The CPD introduces the concept of CE (Communauté Européenne) marking for all construction products. It is therefore essential that those involved in construction are aware of the implications of CE marking. The Directive covers all materials that are permanently incorporated into construction works such as buildings and civil engineering structures. Products can only be used if they can be shown to have the necessary characteristics to meet the six Essential Requirements, (which deal with public safety), as given in the CPD. The CE mark, as such, is not mandatory in the UK.

Further details may be obtained from the European Commission website (see Appendix).

6.3.13 General

Watts and Partners (Chartered Surveyors) have identified more than 135 Acts of Parliament and Regulations applicable to the construction industry with similar lists for Scotland and Northern Ireland. For full lists see *Watts Pocket Book* (Watts 2009).

In addition to the above regulatory provisions it must not be forgotten that several species of creature are protected by law. These include hedgehogs and bats. It is essential that these aspects are considered in redeveloping existing buildings and external facilities.

In addition to the legal restraints described above, practitioners should also be aware that legislation either exists or is envisaged to cover topics such as Arbitration; Construction; White finger; Primary aggregate use; Climate change and Secure and sustainable buildings.

Bibliography and further reference

Adriaanse J. 2005. *Construction contract law: the essentials* Palgrave Macmillan.

Carpenter, J. 2007. Making the most of an opportunity CDM 2007, *The Structural Engineer*, 85(15), IStructE, London.

DETR/ODPM. 1998. *The Party Wall etc. Act. 1996*, explanatory booklet. Published in conjunction with the Welsh Office. (Available free.)
   Note: For further information contact The Faculty of Party Wall Surveyors: Tel: 01424-883-300: email: enq@pws.org.uk.

Doran, D.K. 2007. CDM Regulations – will they make a difference?, Report, *The Structural Engineer*, 85(14), IStructE, London.

ICE. 2007. *Construction law handbook 2007*, 85(14).

Tietz, S. 2007. CDM 2007: some issues for discussion. *The Structural Engineer*, 85(15), IStructE, London.

Watts Group plc. *Watts Pocket Book 2009*, RICS Books, Coventry.

7 Case Studies

7.1 Introduction

Many refurbishment projects have been fully written up in the journals of the relevant professional institutions. An examination of these is worthwhile for a number of reasons that include:

- Gaining knowledge of contemporary construction methods. For example, the paper on the Newport Transporter Bridge gives an early example of the use of structural steel.
- Specific dimensional and other information of structures and buildings.
- Details of investigation of defects including non-destructive and invasive testing.

7.2 The refurbishment of the Newport Transporter Bridge

Opened in 1906 the original design of this 196.6 m clear span steel structure was by French engineer Ferdinand Arnodin. The structure is essentially an aerial ferry (see Figs. 7.1 and 7.2). It was closed in 1985 because of concern regarding its condition and safety. It has a Grade II* listed building status and in the early 1990s sufficient funds became available to carry out a comprehensive refurbishment. A £3m package was put together with contributions from Gwent County Council, European Regional Development Fund, Cadw (Welsh Historic Monuments), Welsh Development Agency and the European Architectural Heritage Fund. The preliminary investigation included a detailed structural survey of every component and every effort was made to match Arnodin's understanding of the bridge behaviour. NDT was used to assist in the initial investigation of the structure. The final cost of the refurbishment was in the order of £3.5m.

The contract was carried out in three phases using standard ICE 5th Edition Conditions of Contract. The contract period was from July 1992 to December 1995.

Engineers: Gwent Consultancy
Contractor: Gwent County Bridge Department
Phase 1: Towers, access stairs and walkways (July 1992 to November 1993).
Phase 2: Main suspension and anchor cable replacement and cable anchorage repairs (December 1993 to December 1994).
Phase 3: Main boom repairs, saddle refurbishment and outstanding works. (June 1995 to December 1995).

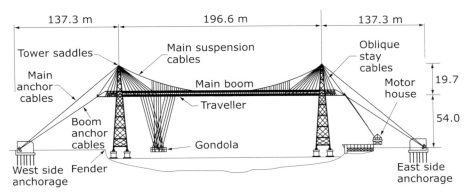

Figure 7.1 Newport Transporter Bridge: general arrangement (Courtesy Barry Mawson).

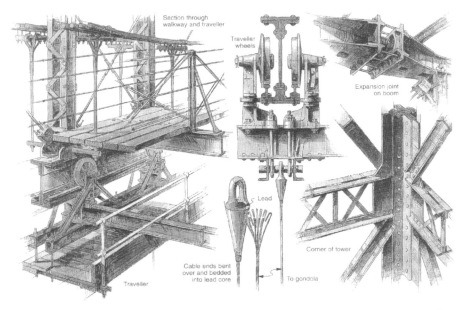

Figure 7.2 Newport Transporter Bridge: miscellaneous details (Courtesy Barry Mawson).

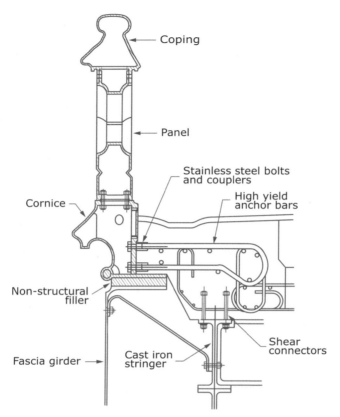

Figure 7.3 Westminster Bridge: details of parapet (Courtesy David Yeoell).

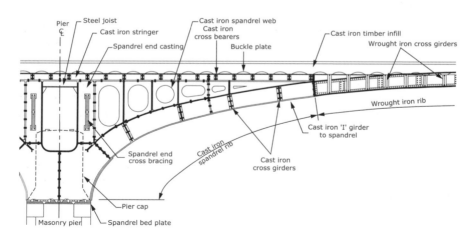

Figure 7.4 Westminster Bridge: details of main rib (Courtesy David Yeoell).

7.3 The refurbishment of Westminster Bridge, London

This 1862 bridge, which is listed Grade II* (i.e. of more than special interest), is part of a complex transport system in which Thames crossings are of considerable importance. (See Figs 7.3 and 7.4.) The maintenance of these structures is essential to ensure that they continue in service and reliably carry modern traffic loading. A principal inspection by Rendell Palmer and Tritton/High-Point in 1989 revealed that it was suffering from many durability problems, mainly caused by water leakage through the deck. There was also concern that the 15-tonne weight limit was being exceeded, causing the buckle plates to be overstressed. Before final decisions on refurbishment were made the City Council decided to carry out a D of T load assessment in accordance with Technical Memorandum BD 21/84. These tests together with strength analyses of cast and wrought iron allowed a complete condition assessment to be made. As a result it was decided to strengthen the structure and to lay a new deck structure of reinforced lightweight aggregate concrete. The work was completed between 1994 and 1997 for a cost in the order of £12m, under main contractor J Murphy and Sons. The original contract was based on the ICE 6th Edition and CESMM3. Because of the nature of the structure many authorities, including English Heritage, The Port of London Authority. Westminster City Council and The Environment Agency were involved putting partnership arrangements under considerable strain. The contractor took the ultimate risks but relied heavily on the integrity and proficiency of others involved. Not least of the problems was to maintain traffic flow across the bridge at all times.

7.4 Windsor Castle – fire behaviour and restoration aspects of historic brickwork

The construction of Windsor Castle dates back to the 11th Century. It has been extended and remodelled several times since then. On 20 November 1992 it was badly damaged by a fire that burned for 15 hours and reached temperatures in excess of 820°C and possibly a lot higher. It is reported that 1.5 Mgal of water was used to quench the blaze. The resulting fire and water damage affected 105 rooms including nine state rooms. An idea of the extent and ferocity of the fire can be gained from Fig. 7.5 which shows a plan of the Brunswick Tower (see also Fig. 7.6).

In the immediate aftermath, the cost of repairing the damage was estimated to be £60m: in the event with skilful professional assistance the cost was limited to £36.5m and the restoration was complete by the Spring of 1998.

Much of the metalwork of the buildings was ironwork (cast and wrought) and dated back to the 1820s, and thought, at the time, to be fireproof. It is suggested that temperatures may have reached 1200°C as the fire in the Brunswick Tower

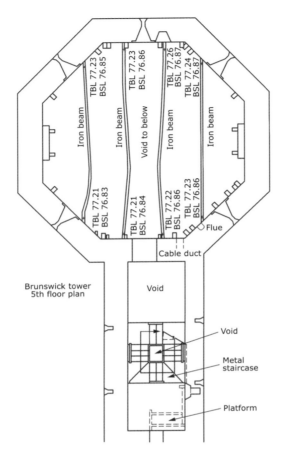

Brunswick tower
5th floor plan

Figure 7.5 Windsor Castle: Brunswick Tower (Courtesy David Dibb-Fuller).

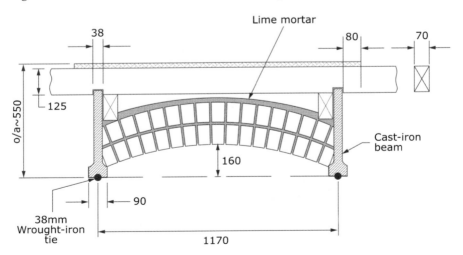

Figure 7.6 Windsor Castle: masonry brick jack arches (Courtesy David Dibb-Fuller).

was allowed to burn itself out. Much of the structure had to be replaced. Testing included metallographic and hardness testing on cast and wrought iron samples.

The restoration was carried out in four phases using a number of consultants and contractors including:

Architects: Bowyer Langlands; Donald Insall & Associates; Sidall Gibson
Engineers: Gifford & Partners; Hockley & Dawson
Services: Dawson partners
Laboratory Testing: Sandberg
Contractors: Wallis Construction; Higgs & Hill

7.5 Listed Georgian Terrace: Manchester Street, London

The project involved the restoration of a terrace of Grade II listed buildings dating from 1790, including the complete reconstruction of two of the buildings. The buildings were returned to their original use as single-family dwellings (see Fig. 7.7).

The 1774 Building Act controlled the design of this terrace, determined how they were constructed and classified them as second rate buildings with a maximum construction cost of £850. The form of construction was therefore known at the outset of the work. Extensive work was required to change the buildings from multiple uses to single family dwellings. In addition to planning permission for the proposed changes Listed Building Consent was also required due to them being listed Grade II. Throughout the works any additional alteration to the fabric required consultation and permission from English Heritage.

Figure 7.7 Listed Georgian terrace, Manchester Street, London (Courtesy Loftus Family Property).

7.5.1 Construction

Each building is basically a brick box with a timber structure inside. Timber floors joists run between the front and rear walls, supported at mid point by a timber stud wall. Openings in the external walls have timber lintels internally and brick arches externally. At roof level, behind the brick parapets are pitched roofs with a timber structure and slate coverings. Gutters and flat areas of roof are made of lead. At the outside walls the floor joists are built into the brickwork. In some locations the joist ends are supported on timber plates built into the brickwork.

7.5.2 Sequence of work

The preservation of certain elements of the fabric had a determining effect on the programme of work. All original joinery and the fireplaces were removed at the outset of the project to enable inspection and, where appropriate, restoration. The original plaster cornices were retained. To avoid the risk of these collapsing, they were stabilised and reinforced before any operation that caused vibration could take place. This was achieved by inserting stainless steel wires into the top of the cornice and tying the wires to the timber joists.

7.5.3 Structural stability

Ensuring structural stability is a key item in refurbishment. Over their 200-year lifespan settlement has occurred at different rates and the soft lime mortar has allowed the brickwork to move to accommodate the changes. In some areas cracks had developed in the brickwork as well as separation of the outside walls from the party walls.

The brick walls are built directly onto the ground. About 600 mm below the basement floor level the width of the brick wall is increased by a series of brick spreaders to reduce the loading on the clay spoil. Defective areas were identified and repaired by a number of techniques. At the junctions between party walls and external walls, angle binders were cast in slots cut into the brickwork to form a strong mechanical link between the walls. Some severe cracks were bridged over with concrete lintels and brickwork cut out and repaired. Extensive use was made of galvanised metal straps to tie floors to walls.

7.5.4 Weatherproofing

The roof structures were basically sound but the slate and lead work had to be stripped and renewed due to their age and condition.

To drain the front parapet gutter and the valley gutter an open gutter passed through the roof space to discharge at the rear of the building. The original was a timber box lined with lead. This was replaced with a sealed 100 mm diameter pipe to reduce the risk of water penetration. The original lead gutter and lead roof used larger sheets and less falls than that suggested by current design practice. This required the design and construction of the gutters to be carried out with great care to retain the existing parapet heights.

7.5.5 Brick walls

The external walls are made of London stock brickwork. Better quality bricks are used on the front walls than on the rear. The party walls use the lowest quality bricks including those that are under burnt, over burnt or misshaped. Overall the brickwork has 4 brick courses to the foot, however for the original ground and first floor levels on the front elevation a more careful selection of bricks gave 4 courses to 11 ¾ inches. The flatness of the original wall also varied. The tolerance on the front elevation was generally 2 mm in 2 m and on the rear elevation 5 mm in 1 m.

Brick repair work and new external walls were constructed from second hand London stocks. Bricks for the front wall were carefully selected and laid to the tight tolerances. The brick arches are constructed from shaped bricks with a flat top and bottom edge.

7.5.6 Rebuilding two properties

The aim of the project was the restoration of the five original houses. Due to a series of events it became necessary to completely rebuild two of the properties. The decision was to use the same construction methods that were used in the original terrace. Solid load-bearing brick walls support the timber floors and roof.

Building Control required these two houses to comply with current regulations for structural stability and to achieve this some modern techniques were required. Concrete foundations were used to support the new brick walls and the original party walls were stabilised using mini piles and ground beams. The brickwork to the party walls was of poor quality and required rendering and grouting.

As the buildings are five stories high (including the basement) the building regulations, at that time, required protection against progressive collapse of the structure. This was achieved with little visual effect on the buildings by casting a reinforced concrete ring beam within the depth of the external brick walls and a concealed metal structure internally.

7.5.7 Timber decay

The thickness of the solid brick walls varies from two and a half bricks thick in the basement to one brick thick (230 mm) on the upper floor. The stock brick is fairly absorbent and any timber built into the wall depth is vulnerable to dampness and decay. All joist ends were inspected for decay and any bond timbers built into the brickwork were removed. Most of these were visible on the inner surface of the wall and would normally occur under joist ends. On some occasions timbers would be within the depth of the brickwork and could only be identified if decay had lead to fault brickwork on the inner surface. Decaying timbers were removed from the roof and all the flat roof and gutter boarding replaced.

7.5.8 Authentic historic materials

A number of historically correct materials were specified for the work including lime mortar, lime putty for brick arches, lime wash, stucco and chimney flaunchings. Good quality materials were readily available. Unless craftsmen have experience in using these materials they will not be able to achieve proper results regardless of their skills. The decision was made to change to modern bagged lime to achieve the required quality.

7.5.9 Windows

The windows are vertical sliding sashes with cast iron weights. They are set back from the outer face of the wall. There was no evidence to suggest any of the windows were original and due to their condition all were replaced. Glazing bars copied the same pattern and shape as the most authentic existing windows. In order to maintain the appearance it was decided to use single glazing but to improve thermal performance by adding a discrete draught stripping. Trickle ventilation was achieved by a simple adaptation of the traditional window head.

All windows on the front elevation had timber shutters. It was possible to restore and re-use some of the original units. To achieve the wall depth needed to house the shutters when open, the original rooms had timber studwork built inside the brickwork.

7.5.10 Thermal insulation

To improve the thermal performance of the buildings the inner face of the brickwork walls had insulation applied. In most cases this was achieved by using fibreglass insulation within the depth of the timber stud wall used to house the window shutters. In other locations insulated plasterboard was applied to the rendered inner face of the wall.[1]

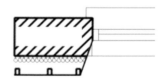

Figure 7.8 Thermal insulation.

- Client: Loftus family Property
- Architect: Hunt Thompson Associates
- Structural Engineer: Ellis Moore
- Services Engineer: Garry Banger
- Consultancy Quantity Surveyor: Axtell Yates Hallet
- Contractor: Ashby & Horner
- Professional Advisor: Dan Cruickshank

1   Original text and illustrations for this case study provided by Richard Pratley.

Figure 7.9 General Register House: Princes Street, Edinburgh (Courtesy Adrian Welch).

7.6 General Register House, Edinburgh: a Grade A listed building

This structure is a Grade A listed building of significant architectural and historic importance. The first section, designed by Robert Adam, was completed in 1788. Between 1822 and 1834 the quadrangle, designed by Robert Reid, was added.

It is the only building in Europe of its age still being used for its original purpose, that is, as a public archive. At the front of this major landmark, at the east end of Princes Street is the bronze, life-size statue of Wellington on his horse. The works comprise a major refurbishment and repair programme consisting of:

- External fabric remedial work such as renewal of slate and lead roof coverings, renewal of rainwater disposal and drainage systems and major re-pointing and repair of ashlar stonework.
- Internal fabric remedial works such as damp proofing of basement areas, localised treatment of timber decay in roof structures, general repairs, cleaning out and redecoration.
- Disabled access/egress provision
- Comprehensive renewal of electrical and mechanical services
- Alterations to space planning for greater layout efficiency

The contract was carried out under a JCT98 Standard (with quantities) Form of Contract; the approximate cost was £18m; the contract was executed in several phases in the period February 1996 to December 2006, including forensic investigation, space planning and design.[2] (See Fig 7.10 for schematic plan.)

---

2 Original text for this case study provided by James Douglas.

Case study: refurbishment and repair of General Register House, Edinburgh

Type of contract: JCT98 Standard Form of Contract (with Quantities)

Key Dates:

- February 1996: report on condition and study of options for repair, upgrading and space planning.
- January 2000: start of Phase One refurbishment and repair project (east–west chosen as the optimal phasing of the work).
- December 2001: completion of Phase One.
- January 2002: start of Phase Two refurbishment and repair project (east–west phasing).
- December 2004: completion of Phase Two.
- January 2005: start of (final) Phase Three refurbishment and repair project (east–west phasing).
- December 2008: completion of Phase Three (delayed due to bankruptcy of original contractor).

Tender value: c.£21,000,000

Consultants

- Project Manager: Turner Townsend Project Management
- Conservation Architect: Gray, Marshall & Associates
- Services Engineer: Blyth & Blyth Associates
- Quantity Surveyor: Thomas & Adamson
- Fire Consultant: Dr Eric Marchant, Edinburgh Fire Consultants Ltd
- Structural Engineer: Wren & Bell
- Impulse Radar Consultants: GB Geotechnics Ltd

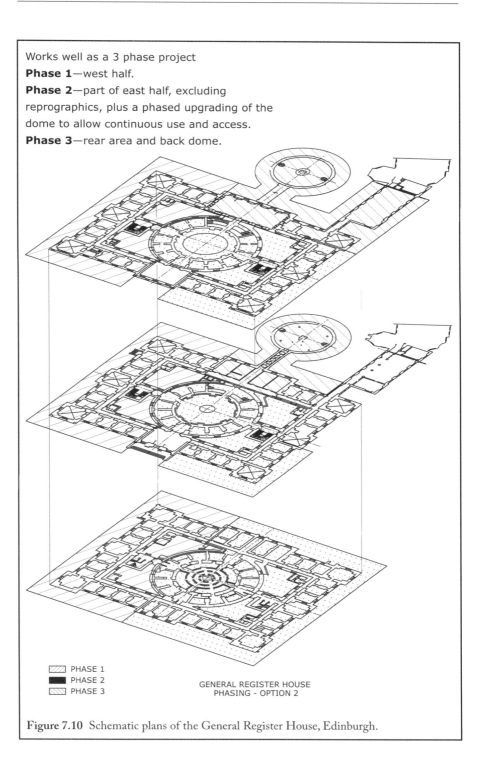

Works well as a 3 phase project
Phase 1—west half.
Phase 2—part of east half, excluding reprographics, plus a phased upgrading of the dome to allow continuous use and access.
Phase 3—rear area and back dome.

PHASE 1
PHASE 2
PHASE 3

GENERAL REGISTER HOUSE
PHASING - OPTION 2

Figure 7.10 Schematic plans of the General Register House, Edinburgh.

7.7 Scottish tenements: a case study in structural repair

This paper by Professor Ian McLeod summarises the outcome of a one-day seminar on the subject in 1984. Scottish tenement blocks were typically built in the late 1800s and were the product of well tried solutions and well tried rule-of-thumb methods. Rarely if ever were they subjected to the rigours of critical structural analysis. They were cheaply built and rarely maintained and, as a result, suffer from many defects including inadequate foundations, lack of horizontal restraint and poor workmanship. The paper also contains several helpful construction diagrams and details of repairs.

The paper gives advice *inter alia* on;

- Structural state of buildings
- Structural advice on alterations
- Inspection of structural matters during reconstruction

See also Section 5.3.2 in Chapter 5.

7.8 Abbreviated case studies

In June 2003 *The Structural Engineer* featured abbreviated studies as follows:

7.8.1 Clifton Suspension Bridge: a Grade 1 listed building

In 2002 12 huge vaults were discovered in the 33 m high Leigh Wood abutment previously thought to be of solid stone and supporting a 26 m high tower. Work to safeguard the future of the bridge has been carried out with the approval of English Heritage and the National Trust. Work carried out by Falcon Structural Repairs and supervised by consulting engineers Nimbus Conservation Ltd.

Figure 7.11 Clifton Suspension Bridge (Courtesy Falcon).

7.8.2 Grand Palais de Paris: innovation in refurbishment techniques

This is an 1897 building on timber piles adjacent to the River Seine on oak piles affected by draw down of river water level. Such piles only remain valid if they remain submerged. Refurbishment work includes the injection of high pressure cement into the foundations, replacement of 20,000 (forged on site) rivets in the metallic superstructure and the replacement of 12,800 m² of glazing. The work has been entrusted to Soletanche Bachy, Eiffel and Dutemple.

7.8.3 The Harley Gallery, Welbeck, Nr Worksop: a small museum with undercroft

Refurbishment includes renovation and stabilisation work. (Consultant Ellis and Brown.)

7.8.4 St Mary's Church, Colton Basset: a 13th century roofless structure

As part of the refurbishment work, the lintels of the tracery windows were reinforced and protected by a reinforced concrete capping beam formed within the rubble-filled cavity stone wall. (Consultant Elliott and Brown.)

7.8.5 Pavilion Lake, Buxton

The original puddle clay liner to this lake was replaced by welded butyl liner. (Consultant Elliott and Brown.)

7.8.6 Latimer's House, Thurcaston, Leicestershire: a 14th century thatched and timbered cottage

This is an example where earlier repair work did not properly address all problem areas. A defective drain had not been repaired causing settlement of the corner of the building. In the second round of repairs it was found necessary to underpin this section and also replace some brickwork with more sympathetic and properly sourced material. The opportunity was also taken to partially reconstruct some of the timber framing. (Consultant: Elliott and Brown.)

Figure 7.12 Latimer's House, Thurcaston, Leicestershire (Courtesy Elliott and Brown).

7.8.7 Stragglethorpe Hall, Lincolnshire: a 17th century half-timbered building

Underwent conversion to a conference centre. Fig. 7.12 shows details of long span sagging timber beams strengthened using steel plates buried within the timber sections. Timber loggia columns whose bases had deteriorated were augmented by new galvanised steel plates sitting on new concrete pads. (Consultant Ellis and Brown.)

7.8.8 The Ikon Gallery, Nr Birmingham: a Grade II listed building

Used as a gallery of contemporary art, the building, built in 1877, was previously a school. It had suffered from a number of defects resulting from fire, dry rot, and eaves spread. A gable was rebuilt, incorporating masonry reinforcement. Other cracked brickwork was stitched and rotting timber replaced. To accommodate increased loading, a discrete new steel frame was inserted inside the building. (Consultant: Steve Evans of Peel and Fowler.)

7.8.9 Queen Elizabeth Hospital Birmingham: a 1930s building incorporating ribbed floors within a steel framed building with brick cladding built tightly around the steelwork

The repair of external masonry required emptying two of four water tanks housed in the seven storey high Clock Tower to allow temporary scaffolding to be put in place. Because of hospital restrictions on dust and noise, severe restrictions were imposed on the masonry repair. (Consultant Steve Evans of Peel and Fowler.)

7.8.10 The Redhouse Cone: a 1790s structure, used until 1939 to house a glass-making furnace

The project involved the refurbishment of the cone itself and the surrounding buildings which date from the 18th and 19th centuries, as well as the installation of new visitor facilities (see Fig. 7.13). Some earlier renovation had been poorly executed and had accelerated low level corrosion. The scheme received a 2003 Structural Heritage Award in 2003. (Consultant Steve Evans of Peel and Fowler.)

Figure 7.13 The Redhouse Cone (Courtesy Steve Evans of Peel and Fowler).

7.8.11 Birmingham School of Jewellery

Victorian and Edwardian buildings built in the period 1860 to 1911. Grade II Listed buildings subjected to bomb blast during the Second World War. The School was established in 1890 but was subjected to a complete renovation in 1994 to provide inspirational surroundings for the design and production of jewellery. Some restoration was required to repair latent Second World War bomb damage which had caused severe water penetration to brickwork containing high levels of sulfates. This included some rebuilding using reclaimed bricks (see Section 5.2.2.9). New atrium walkways supported on cantilever beams were constructed. The project received a RIBA Regional Award and a Civic Trust Award in 1996. (Consultant Steve Evans of Peel and Fowler.)

7.9 Other case studies

7.9.1 Road bridges

- Blakelock R., Munson S.R. and Yeoll D. 1998. The refurbishment of Westminster Bridge: assessment, design and pre-contract planning, *The Structural Engineer*, 76(10), IStructE, London.
- Crossin J., Marshall G.R.D. and Yeoll D. 1998. The refurbishment of Westminster Bridge; bridge strengthening, *The Structural Engineer*, 76(10), IStructE, London.
- Yeoll D., Prasam T. and Hodgkinson B. 2006. Structural repair of elevated Harrow Road (westbound), UK. *Proceedings of ICE, Bridge Engineering*, Thomas Telford, London.
  Note: Contains information about a trial using the desalination technique.
- De Voy J. and Williams J.M. 2007. Strengthening Coalport Bridge, *Structural Engineer International*, 2/2007.

7.9.2 Railway Bridge

- Bessant G.T. 2002. Putney railway bridge. *Proceedings of ICE, Transport*, 153(4), Thomas Telford, London.

7.9.3 Gas Field

- Brown C. 1999. *The Morecombe Bay Gas Field.* In Doran D.K. (Ed.) 1999. *Eminent civil engineers,* Whittles Publishing, Scotland.

7.9.4 Office Block

- Winfield P. *et al.* 1991. The refurbishment of a 1960s office block, *The Structural Engineer*, 69(9), IStructE, London.

7.9.5 Victorian Housing

- Yates. Tim. 2006. *Refurbishing Victorian housing: guidance and assessment.* BRE Information Paper. 1P9/06, BRE, Garston, UK.

7.9.6 Department Store

- Hatter K. 1989. Refurbishment of Whiteleys of Bayswater, *The Arup Journal,* 24(3), Arup, London.

7.9.7 Glass structure

- Jones C. 1987. Restoration of the Palm House, Kew. In: Sandberg, A. (Ed.) *Rehabilitation and renovation*, Proceedings of the 1987 Henderson Colloquium, organised by the British Group of the IABSE, London.

Bibliography and further reference

Blakelock R., Munson S.R. and Yeoll D. 1998. The refurbishment of Westminster Bridge: assessment, design and precontract planning, *The Structural Engineer*, 76/(10), IStructE, London.

Crossin J., Marshall G.R.D. and Yeoll D. 1998. The refurbishment of Westminster Bridge; bridge strengthening. *The Structural Engineer,* 76(10), IStructE, London.

Dibb-Fuller D., Fewtrell R. and Swift R. 1998. Windsor Castle: fire behaviour and restoration aspects of historic ironwork. *The Structural Engineer*, 76(19), IStructE, London.

HMSO. 1984. *The assessment of highway bridges and structures* BD 21/84, HMSO, London (now superseded).

HMSO. 1988. *Loads for highway bridges.* BD37/88, HMSO, London (partly superseded).

Lark R.J. 1966. *An investigation of the suspension system of the Newport Transporter Bridge. Structural assessment – the role of large and full-scale testing,* IStructE/City University.

Lark R.J., Mawson B.R. and Smith A.K. 1999. The refurbishment of Newport Transporter Bridge, *The Structural Engineer*, 77(16), IStructE, London. **See also:** *The Structural Engineer*. 2000. Discussion, *The Structural Engineer*, 78(20), IStructE, London.

MacLeod, I.A. 1989. Scottish tenements: a case study in structural repair, The Structural Engineer, 67(2), IStructE, London.

Pugsley A. and Cullimore M.S.G. 1961 *Report of an investigation of failures of wires in the anchorage cables of the Newport Transporter Bridge,* Bristol University, Bristol.

The Engineer. 1906. The Transporter Bridge at Newport, *The Engineer,* 14 September 1906, Centaur Media, London.

Appendix

A1 Example Gantt Chart

Project: medium-scale refurbishment of a three storey office block

Note: Dark bars represent predicted time-scale; lighter equal actual time-scale (based on Douglas 2006).

Activities	Weeks															
	1	2	3	4	5	6	7	8	9	10	11	12	13	14	15	16
Preparation of site	■/▨															
Preparation of building		■/▨	■/▨													
Basement tanking			■/▨	■/▨	▨											
Over-roofing scheme				■/▨	■/▨	■/▨	■/▨	■/▨	▨							
Over-cladding and new windows						■/▨	■/▨	■/▨	■/▨	■/▨	■/▨	■/▨	■/▨	▨		
Interior refit							■/▨	■/▨	■/▨	■/▨	■/▨	■/▨	■/▨			
Re-commissioning														■	▨	
Tidying up and making good														■		▨
Progress meetings	↑			↑			↑				↑					↑
Safety audits		↑		↑		↑		↑		↑		↑		↑		↑
Milestones		1				2								3		4

Note: Up to 12 weeks prior to the commencement of the contract may be required to obtain the necessary statutory approvals.

A2 Standards, acronyms and symbols

A2.1 British Standards

List of organisations holding reference copies

- Basingstoke Reference Library
- Chelmsford Central Library
- Farnborough Reference Library
- Middlesex University
- Poole Reference Library
- Portsmouth Central Library
- Highbury (Portsmouth) College of Technology Library
- University of Portsmouth
- Southampton Central library
- Hampshire County Library
- Woking Library
- Leicester City Council Library
- Staffordshire County Library
- Blackburn Central Library
- Darlington Borough Council
- Preston District Central Library
- South Tyneside Metropolitan Library
- University of Swansea
- Armagh: Southern Education and Library Board
- Belfast Education and Library Board
- ICE and IStructE libraries

Note: BSI members are provided with a yearbook in hard copy and CD format. This lists all BS, European and ISO standards. Check with BSI website for latest information.

A2.2 Abbreviations

AAC	Aerated Autoclaved Concrete
AAR	Aggregate Alkali Reaction
ACM	Asbestos containing material
AEC	Achieving Excellence in Construction
AP	Approved Document
ASR	Alkali-silica Reaction
BPN	Building Preservation Notice (Order)
BREEAM	Building Research Association Assessment Method
CCA	Copper Chrome Arsenic
CI	Cast iron

C of P	Code of Practice
COSSH	Control of Substances Hazardous to Health Regulations
CPD	Continuing Professional Development
D & B	Design and Build
DBO	Design Build and Operate
DBFO	Design Build Finance and Operate
DD	Draft for Development [BSI documentation]
DDA	Disability Discrimination Act
dpc	Damp proof course
ECC	Engineering and Construction Contract
EOI	Engineering Operating Instruction
FE	Fire extinguisher
FRC	Fibre reinforced cement [Canadian]
FRP	Fibre reinforced plastics polymers
ggbs	Ground granulated blast-furnace slag
GIROD	Glued-in Rods for timber
GPR	Ground Penetrating Radar [aka Impulse radar (IR)]
GRC	Glass-fibre reinforced concrete
GRP	Glass reinforced plastics
HAC	High Alumina Cement
HEPA	High Efficiency Particulate Arrestor
ISAT	Initial Surface Absorption Test
LICONS	Low intrusion conservation systems
LMAC	Liquid metal assisted cracking
LME	Liquid metal embrittlement
MLW	Mean low water
NA	National Annexe
NBS	National Building Studies
NDT	Non Destructive Testing
NEC	New Engineering Contract
NSSC	National Structural Steelwork Specification
NVQ	National Vocational Qualification
OPC	Ordinary Portland cement
PD	Published document [a BSI document]
pfa	Pulverised fuel ash
PPE	Personal Protective Equipment
PSA	Property Services Agency
PVC	Polyvinyl chloride
PWS	Party Wall Surveyors
RAAC	Reinforced Autoclaved Aerated Concrete Slabs
RC	Reinforced concrete
RIDDOR	Reporting of Injuries and Dangerous Occurrences Regulations
RSC	Rolled steel channel
RSJ	Rolled steel joist
SFE	Structural Fire Engineering
SRC	Sulfate-resisting cement
TPO	Tree preservation order
TSA	Thaumasite sulfate attack
TWC	Temporary works coordinator

TWD   Temporary works designer
UPV   Ultrasonic Pulse Velocity
uPVC   Un-plasticised polyvinylchloride
WI   Wrought iron

A2.3 Organisations and institutions for further reference, including acronyms and websites

AA	Aluminium Association (European)	www.aluminium.org
ABE	Association of Building Engineers	www.abe.org.uk
ACE	Association for Consultancy and Engineering	www.acenet.co.uk
ACEC	American Council of Engineering Companies	www.acec.org
ACI	American Concrete Institute	www.aci-int.org
ACR	Advisory Committee for Roofwork	www.roofworkadvice.info
AF	Aluminium Federation	www.alfed.org.uk
AFNOR	Association Française de Normalisation	www.afnor.fr
AISC	American Institute of Steel Construction	www.aisc.org
AISI	American Iron and Steel Institute	www.steel.org
AIST	Association for Iron and Steel Technology	www.aist.org
AITC	American Institute of Timber Construction	www.aitc-glulam.org
ANS	Ancient Monuments Society	www.ancientmonumentssociety.org.uk
ANSI	American National Standards Institute	www.ansi.org
ARB	Architects Registration Board	www.arb.org.uk
ARCA	Asbestos removal Contractors Association	www.arca.org.uk
ARCUK	Architects Registration Council of the United Kingdom [now ARB]	
ASCE	American Society of Civil Engineers	www.asce.org
ASTM	International (formerly American Society for Testing and Materials)	www.astm.org
Barbour	Barbour Index	www.barbour.info/construction
	British Architectural Library	www.riba-library.com
BBA	British Board of Agrément	www.bbacerts.co.uk
BCA	British Cement Association	www.cementindustry.co.uk
BCIRA	British Cast Iron Research Association	www.allbusiness.com
BCIS	Building costs information service	www.bcis.co.uk

BCSA	British Constructional Steelwork Association	www.steelconstruction.org
BDA	Brick Development Association	www.brick.org.uk
BEA	British European Airways [now defunct]	
BINDT	British Institute of Non-Destructive Testing	www.bindt.org
BIS	Belgian Institute for Standardisation	www.cat.bin.be
BITA	British Industrial Truck Association	www.bita.org.uk
BMS	British Masonry Society	www.masonry.org.uk
BNFNF	British Precast Concrete Federation	www.britishprecast.org
BoT	Board of Trade	
BPF	British Plastics Federation	www.bpf.co.uk
BPF	British Property Federation	www.bpf.org.uk
BRE	Building Research Establishment	www.bre.co.uk
	BRE Certification Ltd.	www.brecertification.co.uk
BSC	British Safety Council	www.britishsafetycouncil.org
BS EN	British Standard Euronorm	
BSI	British Standards Institution	www.bsigroup.com
BSI	Building Standards Institute	www.buildingstandards.org
BSRIA	Building Services Research and Information Association Ltd.	www.bsria.co.uk
BSSA	British Stainless Steel Association	www.bssa.org.uk
	British Waterways Board	www.britishwaterways.com
	Building Centre (The)	www.buildingcentre.co.uk
BIS	Bureau of Indian Standards	www.bis.org.in
CADW	Welsh Historic Monuments	www.cadw.wales.co.uk
CARE	Conservation Accreditation Register for Engineers	www.istructe.org.uk
CARES	Certification Authority for	www.ukcares.co.uk
C&CA	Cement and Concrete Association [now defunct: current contact through Concrete Centre]	
CBP	Construction best practice	www.cbppp.org.uk/cbppp
CBDG	Concrete Bridge development Group	www.cbdg.org.uk
CC	Concrete Centre	www.concretecentre.com
CDA	Canadian Dam Association	www.cda.ca
CDA	Copper Development Association	www.cda.org.uk Also: www.copperinfo.co.uk
CDM	Construction (Design &	
CECA	Civil Engineering Contractors Association	www.ceca.co.uk
CEN	European Committee for Standardization	www.cen.eu
CERAM	CERAM Building Technology	www.ceram.com

CFA	Construction Fixings Association	www.fixingscfa.co.uk
CIBSE	Chartered Institution of Building Services Engineers	www.cibse.org
CIC	Construction Industry Council	www.cic.org.uk
CII	Construction Industry Institute	www.construction-institute.org
CIMTEC	Centre for Integrated Monitoring Technology	www.cimtec.com
CIOB	Chartered Institute of Building	www.ciob.org.uk
CIRIA	Construction Industry Research and Information Association	www.ciria.org.uk
CITB	Construction Industry Training Board	www.citbni.org.uk
CLASP	Industrialised building system	www.clasp.gov.uk
	Construction Fixings Association	www.britishtools.com
	Construction Skills	www.constructionskills.net
	Contaminated land	www.contaminatedland.co.uk
CPA	Construction Products Association	www.constructionproducts.org.uk
CPDA	Clay Pipe Development Association	www.cpda.co.uk
CORGI	Council of Registered Gas Installers [Replaced by Gas Safe Register]	www.gassaferegister.co.uk
	Corus [Steelmaker]	www.corusconstruction.com
CRA	Concrete Repair Association	www.cra.org.uk
CROSS	Confidential Reporting on Structural Safety	www.scoss.org.uk/cross
CS	Concrete Society	www.concrete.org.uk
CSI	Construction Specifications Institute	www.csinet.org
CWCT	Centre for Window and Cladding Technology	www.cwct.co.uk
DCLG	Department of Communities and Local Government	www.communities.gov.uk
Defra	Department for Environment, Food and Rural Affairs	www.defra.gov.uk
DES	Department of Education and Science [now Department of Culture Media and Sport]	www.dfes.gov.uk
	Design Council	www.designcouncil.org.uk
DETR	Department of the Environment, Transport and Regions [now Department for Transport Local Government and the Regions]	www.dft.gov.uk
DIN	German Institute for Standardizat	www.2.din.de
DLR	Docklands Light Railway	www.tfl.gov.uk
DOE	Department of the Environment	
DoT	Department of Trade	
DSA	Danish Standards Association	www.ds.dk
DSA	District Surveyors Association	www.londonbuildingcontrol.org.uk
DTI	Department of Trade & Industry	www.dti.gov.uk
DTLG&R	Department of Town, Local Government & Regions [defunct]	

	Dustmite infestation	www.housedustmite.org
	Environment Agency	www.environment-agency.gov.uk
EERI	Earthquake Engineering Research Institute	www.eeri.org
EH	English Heritage	www.englishheritage.org.uk
EMPA	A Swiss research organization associated with ETH Zurich	www.empa.ch
	Eurocodes	www.eurocodes.co.uk
	Eurocode 2	www.eurocode2.info
ECI	European Construction Institute	www.eci-online.org
ECCE	European Council of Civil Engineers	www.eccenet.org
fib	Federation Internationale du Breton	www.fib-international.org
FIDIC	International Federation of Consulting Engineers	www.fidic.org
FIEC	European Construction Industry Federation	www.fiec.org
	Fire Protection Association	www.thepa.co.uk
FMB	Federation of Master Builders	www.fmb.org.uk
FPS	Federation of Piling Specialists	www.fps.org.uk
FRA	Flat Roofing Alliance	www.fra.org.uk
	Friends of the Earth	www.foe.co.uk
FSA	Finnish Standards Association	www.sfs.fi
FSC	Forest Stewardship Council	www.fsc.org
	Galvanizers Association	www.hdg.org.uk
	Gas Safe Register (formerly CORGI)	www.gassaferegister.co.uk
GGF	Glass and Glazing Federation	www.ggf.org.uk
GLC	Greater London Council	www.fsc.org
GRCA	Glassfibre Reinforced Concrete Association	www.grca.org.uk
	Glue Laminated Timber Association	www.glulam.co.uk
HA	Highways Agency	www.highways.gov.uk
HKIE	Hong Kong Institution of Engineers	www.hkie.org.hk
	HM Land Registry	www.landreg.gov.uk
HMSO	Her Majesty's Stationery Office	www.opsi.gov.uk
HO	Home Office	www.homeoffice.gov.uk
HPA	Health Protection Agency	www.hpa.org.uk
HS	Historic Scotland	www.historic-scotland.gov.uk
HSE	Health and Safety Executive [Note: the Health and Safety Commission (HSC) has been subsumed into the HSE.]	www.hse.gov.uk
IABSE	International Association for Bridge and Structural Engineering	www.iabse.org
IABSE	International Association for Bridge and Structural Engineering British Group	www.*iabse*-uk.org

ICE	Institution of Civil Engineers	www.ice.org.uk
ICES	Institution of Civil Engineering Surveyors	www.ices.org.uk
IChemE	Institution of Chemical Engineers	www.icheme.org
ICOMOS	International Council on Monuments and Sites	www.icomos.org
ICT	Institute of Concrete Technology	www.ictech.org
IEE	Institution of Electrical Engineers	www.iee.org
	Institution of Engineers, Australia [Now Engineers Australia]	www.engineersaustralia.org.au
IEI	Institution of Engineers (India)	www.ieindia.org
IEI	Institution of Engineers of Ireland	www.iei.ie
IES	Institution of Engineers Singapore	www.ies.org.sg
IMBM	Institute of Maintenance and Building Management	www.buildingconservation.com
IMechE	Institution of Mechanical Engineers	www.imeche.org.uk
INSB	Italian National Standards Body	www.uni.com
IOSH	Institution of Occupational Safety and Health	www.iosh.co.uk
IoR	Institute of Roofing	www.instituteofroofing.org
IPENZ	Institution of Professional Engineers New Zealand	www.ipenz.org.nz
IQA	Institute of Quality Assurance	www.iqa.org
IRATA	Industrial Rope Access Trade Association	www.irata.org
IRF	Innovation and Research Focus	www.innovationandresearchfocus.org.uk
ISCARSAH	International Scientific Committee of Structures of Architectural Heritage	http://iscarsah.icomos.org
ISO	International Organization for Standardization	www.iso.org
IStructE	Institution of Structural Engineers	www.istructe.org
ITE	Institute of Transportation Engineers	www.ite.org
JBM	Joint Board of Moderators	www.jbm.org.uk
JCT	Joint Contracts Tribunal	www.jctltd.co.uk
JISC	Japanese Industrial Standards Committee	www.jisc.go.jp
JSCA	Japanese Structural Consultants Association	www.jsca.or.jp
JSCE	Japan Society of Civil Engineers	www.jsce-int.org
LCC	London County Council [now defunct]	
LDA	Lead Development Association	www.ldaint.org
LDSA	London District Surveyors Association	www.londonbuildingcontrol.org.uk
LEEA	Lifting Equipment Engineers Association	www.leea.co.uk
	London Metropolitan Archives	www.cityoflondon.gov.uk

	London Museum	www.museumoflondon.org.uk
LSA	Lead Sheet Association	www.leadsheetassociation.org.uk
LUL	London Underground	www.tfl.gov.uk/tube
	Masonry Institute of America	www.masonryinstitute.org
	Meteorological Office	www.meto.gov.uk
NATM	New Austrian Tunneling Method	
NBS	National Building Specification Ltd.	www.nbsservices.co.uk
NCB	National Coal Board	
	Network Rail	www.networkrail.co.uk
NFDC	National Federation of Demolition Contractors	www.demolition-nfdc.com
NFPA.	National Fire Protection Association	www.nfpa.org
NGRPA	National Glass Reinforced Plastics Association	
NHBC	National House Building Council	www.nhbc.co.uk
NIST	National Institute of Standards and Technology	www.nist.gov
NJUG	National Joint Utilities Group	www.njug.org.uk
NMAB	National Materials Advisory Board	www.nationalacademies.org
NRPB	National Radiological Protection Board	www.nrpb.org
NSC	New Steel Construction	www.new-steel-construction.com
NSI	Netherlands Standards Institute	www.tue.nl
NSC	National Safety Council	www.nsc.org
ODPM	Office of the Deputy Prime Minister [now subsumed into the DCLG]	www.odpm.gov.uk
OGC	Office of Government Commerce	www.ogc.gov.uk
ON	Austrian Standards Institute	www.on-norm.at
OS	Ordnance Survey	www.ordnancesurvey.co.uk
PCA	Portland Cement Association	www.cement.org
	Post-Tensioning Institute	www.post-tensioning.org
PRA	Paint Research Association	www.pra.org.uk
	Precast/Prestressed Concrete Institute	www.pci.org
	Pyramus and Thisbe Club	www.partywalls.org.uk
QPA	Quarry Products Association	www.qpa.org
	Railtrack plc [now Network Rail]	www.networkrail.com
RAPRA	Rubber and Plastics Research Association	www.rapra.net
	Ready Mixed Concrete Bureau	www.rcb.org.uk
RGS	Royal Geological Society	www.britstra.org
RIBA	Royal Institute of British Architects	www.riba.org
RICS	Royal Institution of Chartered Surveyors	www.rics.org
RIL	Association of Finnish Civil Engineers	www.ril.fi
RILEM	International Union of Laboratories and Experts in Construction Materials, Systems and Structures	www.rilem.org
	Royal Engineers Museum	www.remuseum.org.uk
RoSPA	Royal Society for the Prevention of Accidents	www.rospa.com

SAFCEC	South African Federation of Civil Engineering Contractors	www.safcec.org.za
SAICE	South African Institution of Civil Engineering	www.civils.org.za
SAISC	Southern African Institute of Steel Construction	www.saisc.co.za
	Scottish Building Regulations	www.scotland.gov.uk
SCC	Standards Council of Canada	www.scc.ca
SCI	Society of Chemical Industry	www.soci.org
SCI	Steel Construction Institute	www.steel-sci.org
SCOSS	Standing Committee on Structural Safety	www.scoss.org.uk
SDD	Scottish Development Department [Now the Scottish Executive Development Department (SEDD)]	www.scotland.gov.uk
	Specify-it	www.specify-it.com
SESOC	Structural Engineering Society	www.sesoc.org.nz
SIS	Swedish Standards Institute	www.sis.se
SPAB	Society for the Preservation of Ancient Buildings	www.spab.org.uk
	Stone Federation Great Britain	www.stone-federationb-org.uk
SSHA	Scottish Special Housing Association [now defunct]	
TRADA	Timber Research and Development Association	www.trada.co.uk
TRF	Transport Research Federation	www.transportresearchfoundation.co.uk
TRL	Transport Research Laboratory	www.trl.co.uk
TSE	*The Structural Engineer*	www.istructe.org.uk/thestructuralengineer
TSO	The Stationery Office [was HMSO]	www.tso.co.uk
TTF	Timber Trade Federation	www.ttf.co.uk
UKAS	United Kingdom Accreditation Service	www.ukas.com
	Water Authorities Association	www.water.org.uk
	Water UK	www.water.org.uk
WM&R	Water Management and Research Group	www.cf.ac.uk
WRc	Water Research Council	www.wrcplc.co.uk
	Weald and Downland Open Air Museum	www.wealddown.co.uk

A3 Imperial/metric conversions

Conversion Chart (to 3 significant figures)

Measure	Imperial to SI units	SI to imperial units
Length	1 yd = 0.914 m	1 m = 1.09 yd = 3.28 ft
	1 ft = 0.305 m	1 cm = 0.394 in
	1 in = 25.4 mm	1 mm = 0.0394 in
Area	1 yd^2 = 0.836 m^2	1 m^2 = 1.20 yd^2 = 10.8 ft^2
	1 ft^2 = 0.09290 m^2	1 cm^2 = 0.155 in^2
	1 in^2 = 645 mm^2	1 mm^2 = 0.00155 in^2
Volume	1 yd^3 = 0.765 m^3	1 m^3 = 1.31 yd^3 = 35.3 ft^3
	1 ft^3 = 0.0283 m^3	1 cm^3 = 0.0610 in^3
	1 in^3 = 16400 mm^3	1 litre = 0.220 gallons
	1 gallon = 4.55 litres	
Mass	1 ton = 1020 kg = 1.020 tonne	1 tonne = 0.984 ton
	1 cwt = 50.8 kg	1 kg = 2.20 lb
	1 lb = 0.454 kg	
Density	1 lb/ft^3 = 16.0 kg/m^3	1 kg/m^3 = 0.0624 lb/ft^3
Force	1 tonf = 9.96 kN	1 N = 0.225 lbf
	1 lbf = 4.45 N	1 kN = 225 lbf = 0.100 ton
Pressure	1 tonf/ft^2 = 107 kN/m^2	1 kN/m^2 = 0.00932 ton/ft^2
	1 tonf/in^2 =15.4 N/mm^2	1 kN/m^2 = 20.9 lbf/ft^2
	1 lbf/in^2 = 0.00689 N/mm^2	1 N/mm^2 (1MPa) = 145 lbf/in

For more detailed information on conversion from Imperial to SI units and *vice versa* see BS 350: Part 1, 1974. Chart taken from 1 STRUCTE Report 'Appraisal of Existing Structures' 2nd Edition 1996.

A4 Table of atomic symbols

Element	Symbol	Element	Symbol
Actinium	Ac	Berkelium	Bk
Aluminum	Al	Beryllium	Be
Americium	Am	Bismuth	Bi
Antimony	Sb	Boron	B
Argon	Ar	Bromine	Br
Arsenic	As	Cadmium	Cd
Astatine	At	Calcium	Ca
Barium	Ba	Californium	Cf

Element	Symbol	Element	Symbol
Carbon	C	Neodymium	Nd
Cerium	Ce	Neon	Ne
Cesium	Cs	Neptunium	Np
Chlorine	Cl	Nickel	Ni
Chromium	Cr	Niobium	Nb
Cobalt	Co	Nitrogen	N
Copper	Cu	Nobelium	No
Curium	Cm	Osmium	Os
Dysprosium	Dy	Oxygen	O
Einsteinium	Es	Palladium	Pd
Erbium	Er	Phosphorus	P
Europium	Eu	Platinum	Pt
Fermium	Fm	Plutonium	Pu
Fluorine	F	Polonium	Po
Francium	Fr	Potassium	K
Gadolinium	Gd	Praseodymium	Pr
Gallium	Ga	Promethium	Pm
Germanium	Ge	Protactinium	Pa
Gold	Au	Radium	Ra
Hafnium	Hf	Radon	Rn
Helium	He	Rhenium	Re
Holmium	Ho	Rhodium	Rh
Hydrogen	H	Rubidium	Rb
Indium	In	Ruthenium	Ru
Iodine	I	Samarium	Sm
Iridium	Ir	Scandium	Sc
Iron	Fe	Selenium	Se
Krypton	Kr	Silicon	Si
Lanthanum	La	Silver	Ag
Lead	Pb	Sodium	Na
Lithium	Li	Strontium	Sr
Lutetium	Lu	Sulphur	S
Magnesium	Mg	Tantalum	Ta
Manganese	Mn	Technetium	Te
Mendelevium	Md	Tellurium	Te
Mercury	Hg	Terbium	Tb
Molybdenum	Mo	Thallium	Tl

Element	Symbol	Element	Symbol
Thorium	Th	Vanadium	V
Thulium	Tm	Xenon	Xe
Tin	Sn	Ytterbium	Yb
Titanium	Ti	Yttrium	Y
Tungsten	W	Zinc	Zn
Uranium	U	Zirconium	Zr

A5 The Greek Alphabet

Capital	Lower-case	Name	English transliteration
A	α	alpha	a
B	β	beta	b
Γ	γ	gamma	g
Δ	δ	delta	d
E	ε	epsilon	e
Z	ζ	zeta	z
H	η	eta	$\bar{e}$
Θ	θ	theta	th
I	ι	iota	i
K	κ	kappa	k
Λ	λ	lanbda	l
M	μ	mu	m
N	ν	nu	n
Ξ	ξ	xi	x
O	o	moicron	o
Π	π	pi	p
P	ρ	rho	r
Σ	σ (ς at end of word)	sigma	s
T	τ	tau	t
Y	υ	upsilon	u
Φ	φ	phi	ph
X	χ	chi	kh
Ψ	ψ	psi	ps
Ω	ω	omega	$\bar{o}$

A6 Properties of construction timber

Table A6.1 Properties and uses of hardwoods.

Species	Colour (kg m⁻³)	Density	Texture	Moisture movement	Working qualities	Durability	Permeability	Uses
Abura *Mitragyna ciliata* W. Africa	Hardwood Light brown	580*	Medium/ fine	Small	Medium	Non-durable	Moderately resistant	Interior joinery. Mouldings
Afrormosia *Pericopsis elata* W. Africa	Hardwood Light brown	710	Medium/ fine	Small	Medium	Very durable	Extremely resistant	Interior and exterior joinery. Furniture. Cladding
Afzelia/doussié *Afzelia* spp. W. Africa	Hardwood Reddish-brown	830*	Medium/ coarse	Small	Medium/ difficult	Very durable	Extremely resistant	Interior and exterior joinery. Cladding
Agba *Gossweilerodendron balsamiferum* W. Africa	Hardwood Yellowish-brown	510	Medium	Small	Good	Durable	Resistant	Interior and exterior joinery. Cladding
Andiroba *Carapa guianensis* S. America	Hardwood Pink to red-brown	640	Medium/ coarse	Small	Medium	Moderately durable	Extremely resistant	Interior joinery
Ash. American *Fraxinus* spp. USA	Hardwood Grey, brown	670	Coarse	Medium	Medium	Non-durable	Permeable	Interior joinery. Trim. Tool handles
Ash. European *Fraxinus excelsior* Europe	Hardwood White to light brown	710*	Medium/ coarse	Medium	Good	Perishable	Moderately resistant	Interior joinery. Sports goods
Aspen *Populus tremuloides* Canada, USA	Hardwood Grey, white to pale brown	450	Fine	Large	Medium	Perishable/ non-durable	Extremely resistant	Interior joinery. Matches.
Balsa *Ochroma pyramidale* S. America	Hardwood White	160*	Fine	Small	Good	Perishable	Resistant	Useful for heat, sound and vibration insulation. Buoyancy aids.

Table A6.1 Properties and uses of hardwoods *(continued)*

Species	Colour (kg m⁻³)	Density	Texture	Moisture movement	Working qualities	Durability	Permeability	Uses
Balau† *Shorea* spp. S.E. Asia	Hardwood Yellow-brown to red-brown	980	Medium	Medium	Medium	Very durable	Extremely resistant	Heavy structural work, bridge and wharf construction
Balau, Red *Shorea* spp. S.E. Asia	Hardwood Purplish-red or dark red-brown	880	Medium	Medium	Medium	Moderately durable	Extremely resistant	Heavy structural work
Basralocus *Dicorynia guianensis* Surinam, French Guiana	Hardwood Lustrous brown	720	Medium	Medium/large	Medium	Very durable	Extremely resistant	Marine and heavy construction
Basswood *Tilia americana* N.America	Hardwood Creamy white to pale brown	420	Fine	Medium	Good	Non-durable	Permeable	Constructional veneer, turnery, piano keys, woodware
Beech, European *Fagus sylvatica* Europe	Hardwood Whitish to pale brown, pinkish-red when steamed	720	Fine	Large	Good	Perishable	Permeable	Furniture. interior joinery. Flooring. Plywood
Birch, American *Betula* spp. N. America	Hardwood Light to dark reddish-brown	710	Fine	Large	Good	Perishable	Moderately resistant	Furniture. Plywood. Flooring
Birch, European *Betula pubescens* Europe, Scandinavia	Hardwood White to light brown	670	Fine	Large	Good	Perishable	Permeable	Plywood. Furniture. Turnery
Cedar. Central/South American *Cedrela* spp. Central & S. America	Hardwood Pinkish-brown to dark reddish-brown	480	Coarse	Small	Good	Durable	Extremely resistant	Cabinet work, interior joinery. Racing-boat building. Cigar boxes

Table A6.1 Properties and uses of hardwoods *(continued)*

Species	Colour	Density (kg m⁻³)	Texture	Moisture movement	Working qualities	Durability	Permeability	Uses
Cherry, American *Prunus serotina* USA	Hardwood Reddish-brown to red	580	Fine	Medium	Good	Moderately durable	No information	Cabinet making. Furniture. Interior joinery
Cherry, European *Prunus acium* Europe	Hardwood Pinkish-brown	630	Fine	Medium	Good	Moderately durable	No information	Cabinet making. Furniture
Chesnut, horse *Aesculus hippocastanum* Europe	Hardwood White to pale yellow-brown	510	Fine	Small	Medium	Perishable	Permeable	Brush backs. Fruit trays and boxes
Chesnut, sweet *Castanea sativa* Europe	Hardwood Yellowish-brown	560	Medium	Large	Good	Durable	Extremely resistant	Interior and exterior joinery. Fencing
Danta *Nesogordonia papaverifera* W. Africa	Hardwood Yellowish-brown	750	Fine	Medium	Good	Moderately durable	Resistant	Flooring. Joinery. Turnery
Ebony *Diospyros spp.* W. Africa, India Sri Lanka	Hardwood Black, some grey/ black stripes	1030/1190	Fine	Medium	Medium	Very durable	Extremely resistant	Used primarily for decorative work. Turnery. Inlaying
Ekki/azobé¹ *Lophira elata* W. Africa	Hardwood Dark red to dark brown	1070	Coarse	Large	Difficult	Very durable	Extremely resistant	Heavy construction, marine and freshwater construction. Bridges, sleepers, etc.
Elm, American *Ulmus americana* N. America	Hardwood Pale reddish-brown	580	Coarse	Medium	Medium	Non-durable	Moderately resistant	Furniture. Coffins. Rubbing strips
Elm, European *Ulmus spp.* Europe	Hardwood Light brown	560*	Coarse	Medium	Medium	Non-durable	Moderately resistant	Furniture. Coffins. Boat building

Table A6.1 Properties and uses of hardwoods *(continued)*

Species	Colour	Density (kg m⁻³)	Texture	Moisture movement	Working qualities	Durability	Permeability	Uses
Freijo *Cordia goeldiana* S.America	Hardwood Golden brown	590	Medium	Medium/small	Medium	Durable	No information	Furniture. Interior and exterior joinery
Gaboon *Aucoumea klaineana* W. Africa	Hardwood Pinkish-brown	430	Medium	Medium	Medium	Non-durable	Resistant	Used principally for plywood and blockboard
Gedu nohor/edinam *Entandrophragma angolense* W. Aftrica	Hardwood Pinkish-brown	560	Medium	Small	Medium	Moderately durable	Extremely resistant	Furniture. Interior and exterior joinery
Geronggang *Cratoxylon arborescens* S.E. Asia	Hardwood Pink to red	550	Coarse	Medium	Medium	Non-durable	Permeable	Interior joinery
Greenheart† *Ocotea rodiaei* Guyana	Hardwood Yellow/olive green to brown	1040	Fine	Medium	Difficult	Very durable	Extremely resistant	Heavy construction, marine and freshwater construction. Bridges, etc.
Guarea *Guarea cedrata* W. Africa	Hardwood Pinkish-brown	590	Medium	Small	Medium	Very durable	Extremely resistant	Furniture, interior joinery, cabinet making
Hickory *Carya* spp. N. American	Hardwood Brown to reddish-brown	830	Coarse	Large	Difficult	Non-durable	Moderately resistant	Striking tool handles, ladder rungs, sports goods
Idigbo *Terminalia ivorensis* W. Africa	Hardwood Yellow	560*	Medium	Small	Medium	Durable	Extremely resistant	Interior and exterior joinery, plywood
Iroko *Chlorophora excelsa* W. Africa	Hardwood Yellow-brown	660	Medium	Small	Medium/difficult	Very durable	Extremely resistant	Exterior and interior joinery. Bench tops. Constructional work

Table A6.1 Properties and uses of hardwoods (continued)

Species	Colour (kg m⁻³)	Density	Texture	Moisture movement	Working qualities	Durability	Permeability	Uses
Jarrah[†] *Eucalyptus marginata* Australia	Hardwood Pink to dark red	820*	Medium	Medium	Difficult	Very durable	Extremely resistant	Heavy constructional work. Flooring
Jelutong *Dyera costulata* S.E. Asia	Hardwood White to yellow	470	Fine	Small	Good	Non-durable	Permeable	Pattern making. Drawing boards
Karri[†] *Dryobalanops* spp. S.E. Asia	Hardwood Reddish-brown	770*	Medium	Medium	Medium	Very durable	Extremely resistant	Exterior joinery. Decking. Constructional use
Karri[†] *Eucalyptus diversicolor* Australia	Hardwood Reddish-brown	900	Medium	Large	Difficult	Durable	Extremely resistant	Heavy construction
Kauvula *Endospermum macrophyllum* Fiji	Hardwood Pale cream to straw-yellow	480	Medium to coarse	Small	Medium	Perishable	Permeable	Mouldings. Interior joinery
Kempas[†] *Koompassia malaccensis* S.E. Asia	Hardwood Orange-red to red-brown	880	Course	Medium	Difficult	Durable	Resistant	Heavy constructional use
Keruing, apitong, gurjun, yang[†] *Dipterocarpus* spp. S.E. Asia	Hardwood Pinkish-brown to dark brown	740*	Medium	Large/ medium	Difficult	Moderately durable	Resistant	Heavy and general construction. Decking, vehicle flooring
Lauan see Meranti								
Lignum vitae *Guaiacum* spp. Central America	Hardwood Dark green/ brown	1250	Fine	Medium	Difficult	Very durable	Extremely resistant	Bushes and bearings. Sports goods and textile equipment

Table A6.1 Properties and uses of hardwoods *(continued)*

Species	Colour (kg m⁻³)	Density	Texture	Moisture movement	Working qualities	Durability	Permeability	Uses
Limba/afara *Terminalia superba* W. Africa	Hardwood Pale yellow-brown/ straw	560*	Medium	Small	Good	Non-durable	Moderately resistant	Furniture. Interior joinery
Lime, European *Tilia* spp. Europe	Hardwood Yellowish-white to pale brown	560	Fine	Medium	Good	Perishable	Permeable	Carving. Turnery. Bungs. Clogs
Mahogany, Africa *Khaya* spp. W. Africa	Hardwood Reddish-brown	530	Medium	Small	Medium	Moderately durable	Extremely resistant	Furniture. Cabinet work. Boat building. Joinery
Mahogany, American *Swietenia macrophylla* Central and S. America, especially Brazil	Hardwood Reddish-brown	560	Medium	Small	Good	Durable	Extremely resistant	Furniture. Cabinet work. Interior and exterior joinery. Boat building
Makoré *Tieghemella heckelii* W. Africa	Hardwood Pinkish-brown to dark red	640	Fine	Small	Medium	Very durable	Extremely resistant	Furniture. Interior and exterior joinery. Boat building. Plywood
Maple, rock *Acer saccharum* N. America	Harwood Creamy white	740	Fine	Medium	Medium	Non-durable	Resistant	Excellent flooring timber. Furniture. Sports goods
Maple, soft *Acer saccharinum* N. America	Harwood Creamy white	650	Fine	Medium	Medium	Non-durable	Moderately resistant	Furniture. Interior joinery. Turnery
Mengkulang *Heritiera* supp. S.E. Asia	Hardwood Red, brown	720	Coarse	Small	Medium	Moderately durable	Resistant	Interior joinery. Construction. Plywood
Meranti, dark red/ dark red seraya/ red lauan *Shorea* spp. S.E. Asia	Hardwood Medium to dark red-brown	710*	Medium	Small	Medium	Variable, generally moderately durable to durable	Resistant to extremely resistant	Interior and exterior joinery. Plywood

Table A6.1 Properties and uses of hardwoods (continued)

Species	Colour	Density (kg m⁻³)	Texture	Moisture movement	Working qualities	Durability	Permeability	Uses
Meranti, light red/ light red seraya/ white lauan *Shorea* spp. S.E. Asia	Hardwood Pale pink to mid-red	550*	Medium	Small	Medium	Variable, generally non-durable to moderately durable	Extremely resistant	Interior joinery. Plywood
Meranti, yellow/ yellow seraya *Shorea* spp. S.E. Asia	Hardwood Yellow-brown	660*	Medium	Small	Medium	Variable, generally non-durable to moderately durable	Extremely resistant	Interior joinery. Plywood
Merbau† *Intsia* spp. S.E. Asia	Hardwood Medium to dark re-brown	830	Coarse	Small	Moderate	Durable	Extremely resistant	Joinery. Flooring structural work
Nemesu *Shorea pauciflora* Malaysia	Hardwood Red-brown to dark red	710	Medium	Small	Medium	Moderate durable to durable	Resistant to extremely resistant	Interior and exterior joinery. Plywood
Niangon *Tarrietia utils* W. Africa	Hardwood Reddish-brown	640*	Medium	Medium	Good	Moderately durable	Extremely resistant	Interior and exterior joinery. Furniture
Nyatoh *Palaquium* spp. S.E. Asia	Hardwood Pale pink to red-brown	720	Fine	Medium	Medium	Non-durable to moderately durable	Extremely resistant	Interior joinery. Furniture
Oak, American red *Quercus* spp. N. America	Hardwood Yellowish-brown with red tinge	790	Medium	Medium	Medium	Non-durable	Moderately resistant	Furniture. Interior joinery
Oak, American white *Quercus* spp. N. America	Hardwood Pale yellow to mid-brown	770	Medium	Medium	Medium	Durable	Extremely resistant	Furniture. Cabinet work. Tight cooperage

Table A6.1 Properties and uses of hardwoods *(continued)*

Species	Colour	Density (kg m^{-3})	Texture	Moisture movement	Working qualities	Durability	Permeability	Uses
Oak, European *Quercus robur* Europe	Hardwood Yellowish-brown	670/720	Medium/ coarse	Medium	Medium/ difficult	Durable	Extremely resistant	Furniture. Interior and exterior joinery. Flooring. Tight cooperage. Fencing
Oak, Japanese *Quercus mongolica* Japan	Harwood Pale yellow	670	Medium	Medium	Medium	Moderately durable	Extremely resistant	Furniture. Interior joinery
Oak, Tasmanian *Eucalyptus delegatensis Eucalyptus obliqua Eucalyptus regnans* Australia. Tasmania	Hardwood Pale pink to brown	610/710	Coarse	Medium	Medium	Moderately durable	Resistant	Furniture. Interior joinery
Obeche *Triplochiton scleroxylon* W. Africa	Hardwood White to pale yellow	390	Medium	Small	Good	Non-durable	Resistant	Interior joinery. Furniture. Plywood
Opepe† *Nauclea diderrichil* W. Africa	Hardwood Yellow to orange- yellow	750	Coarse	Small	Medium	Very durable	Moderately resistant	Heavy construction work. Marine and freshwater use. Exterior joinery. Flooring
Padauk *Pterocarpus spp.* W. Africa, Andamans, Burma	Hardwood Red to dark purple-brown	640/* 850	Coarse	Small	Medium	Very durable	Moderately resistant to resistant	Interior and exterior joinery. Turnery. Flooring
Pau marfim *Balfourodendron riedelianum* S. America	Hardwood Yellow	800	Medium	Large	Good	Non-durable	No information	Interior joinery. Furniture. Flooring
Plane, European *Platanus hybrida* Europe	Hardwood Mottled red- brown	640	Fine	No information	Medium	Perishable	No information	Decorative purposes. Inlay work

Table A6.1 Properties and uses of hardwoods *(continued)*

Species	Colour	Density (kg m⁻³)	Texture	Moisture movement	Working qualities	Durability	Permeability	Uses
Poplar *Populus* spp. Europe	Hardwood Grey, white to pale brown	450	Fine/ medium	Large	Medium	Perishable non-durable	Extremely resistant	Pallet blocks. Box boards. Turnery. Wood wool
Purpleheart *Peltogyne* spp. Central & S. America	Hardwood Purple to purplish-brown	880	Medium	Small	Medium/ difficult	Very durable	Extremely resistant	Heavy construction. Flooring. Turnery
Ramin *Gonystylus* spp. S.E. Asia	Hardwood White to pale yellow	670	Medium	Large	Medium	Non-durable	Permeable	Mouldings, furniture
Rosewood *Dalbergia* spp. S. America, India	Hardwood Medium to dark purplish-brown with black streaks	870*	Medium	Small	Medium	Very durable	Extremely resistant	Interior joinery. Cabinet work. Turnery
Sapele *Entandrophragma cylindricum* W. Africa	Hardwood Medium reddish-brown with market stripe figure	640	Medium	Medium	Medium	Moderately durable	Resistant	Interior joinery. Furniture. Flooring
Sepetir *Sindora* spp. S.E. Asia	Hardwood Golden brown	640/* 830	Medium	Small	Difficult	Durable	Extremely resistant	Joinery. Furniture
Seraya – see Meranti								
Sycamore *Acer pseudoplatanus* Europe	Hardwood White or yellowish-white	630	Fine	Medium	Good	Perishable	Permeable	Turnery. Textile equipment. Joinery
Taun *Pometia pinnata* S.E. Asia	Hardwood Pale pinkish-brown	750	Coarse	Medium	Medium	Moderately durable	Moderately resistant	Structural work. Turnery. Joinery. Furniture

Table A6.1 Properties and uses of hardwoods *(continued)*

Species	Colour (kg m⁻³)	Density	Texture	Moisture movement	Working qualities	Durability	Permeability	Uses
Teak¹ *Tectona grandis* Burma, Thailand	Hardwood Golden brown, sometimes with dark markings	660	Medium	Small	Medium	Very durable	Extremely resistant	Furniture. Interior and exterior joinery. Boat building
Utile *Entandrophragma utile* W. Africa	Hardwood Reddish-brown	660	Medium	Medium	Medium	Durable	Extremely resistant	Interior and exterior joinery. Furniture and cabinet work
Virola/baboen *Virola* spp. *Dialyanthera* supp. S. America	Hardwood Pale pinkish-brown	430/* 670	Medium	Medium	Medium	Non-durable	Permeable	Carpentry. Furniture. Plywood. Moulding
Wallaba *Eperua falcata Eperua grandiflora* Guyana	Hardwood Dull reddish-brown	910	Coarse	Medium	Medium	Very durable	Extremely resistant	Transmission poles. Flooring. Decking. Heavy construction
Walnut, Africa *Lavoa trichilioides* W. Africa	Hardwood Yellowish-brown sometimes with dark streaks	560	Medium	Small	medium	Moderately durable	Extremely resistant	Furniture. Cabinet work. Interior and exterior joinery
Walnut, America *Juglans nigra* N. America	Hardwood Rich dark brown	660	Coarse	Small/ medium	Good	Very durable	Resistant	Furniture. Gun stocks
Walnut, European *Juglans regia* Europe	Hardwood Grey-brown with dark streaks	670	Coarse	Medium	Good	Moderately durable	Resistant	Furniture. Turnery. Gun stocks
Wenge *Millettia laurentii Millettia stuhlmannii* Central & E. Africa	Hardwood Dark brown with fine black veining	880	Coarse	Small	Good	Durable	Extremely resistant	Interior and exterior joinery. Flooring. Turnery

Table A6.1 Properties and uses of hardwoods *(continued)*

Species	Colour (kg m⁻³)	Density	Texture	Moisture movement	Working qualities	Durability	Permeability	Uses
Willow *Salix* spp. Europe	Hardwood Pinkish-white	450	Fine	Small	Good	Perishable	Resistant	Cricket bats. Boxes. Crates

* Density can vary by 20% or more
† Structural properties included in BS 5268: Part 2: 1988

Table A6.2 Properties and uses of softwoods.

Species	Colour (kg m⁻³)	Density	Texture	Moisture movement	Working qualities	Durability	Permeability	Uses
Cedar of Lebanon *Cedrus libani* Europe	Softwood Light brown	580	Medium	Medium/small	Good	Durable	Resistant	Joinery. Garden furniture. Gates
Douglas Fir† *Pseudotsuga menziesii* N. America and UK	Softwood Light reddish-brown	530	Medium	Small	Good	Moderately durable	Resistant/extremely resistant	Plywood. Interior and exterior joinery. Construction. Vats and tanks
Hemlock, Western† *Tsuga heterophylla* N. America	Softwood	500	Fine	Small	Good	Non-durable	Resistant	Construction. Joinery
Larch, European† *Larix decidua* Europe	Softwood Pale reddish-brown	590	Fine	Small	Medium	Moderately durable	Resistant	Boat planking. Pit props. Transmission poles
Larch, Japanese† *Larix kaempferi* Europe	Softwood Reddish-brown	560	Fine	Small	Medium	Moderately durable	Resistant	Stakes. General construction

Table A6.2 Properties and uses of softwoods *(continued)*

Species	Colour Density (kg m⁻³)	Density	Texture	Moisture movement	Working qualities	Durability	Permeability	Uses
Parana pine[†] *Araucaria angustifolia* S. America	Softwood Golden-brown with bright red streaks	550	Fine	Medium	Good	Non-durable	Moderately resistant	Interior joinery. Plywood
Pine, Corsican[†] *Pinus nigra* Europe	Softwood Light yellowish-brown	510	Coarse	Small	Medium	Non-durable	Moderately resistant	Joinery. Construction
Pine, Maritime *Pinus pinaster* Europe	Softwood Pale brown to yellow	510	Medium/coarse	Medium	Good	Moderately durable	Resistant	Pallets and packaging
Pine, pitch[†] *Pinus palustris Pinus elliottii* Southern, USA	Softwood Yellow-brown to red-brown	670	Medium	Medium	Medium	Moderately durable	Resistant	Interior and exterior joinery. Heavy construction
Pine, radiata *Pinus radiata* S. Africa, Australia	Softwood Yellow to pale brown	480	Medium	Medium	Good	Non-durable	Permeable	Furniture. Packaging
Pine, Scots[†] *Pinus sylvestris* UK	Softwood Pale yellowish-brown to red-brown	510	Coarse	Medium	Medium	Non-durable	Moderately resistant	Construction. Joinery
Pine, Southern[†] A number of species including *Pinus palustris, Pinus elliottii, Pinus echinata, Pinus taeda* Southern USA	Softwood Pale yellow to light brown	560*	Medium	Medium	Medium	Non-durable	Moderately resistant	Construction. Joinery. Plywood
Pine, yellow *Pinus strobus* N. America	Softwood Pale yellow to light brown	420	Fine	Small	Good	Non-durable	Moderately resistant	Pattern making. Drawing Boards. Doors

Table A6.2 Properties and uses of softwoods *(continued)*

Species	Colour (kg m⁻³)	Density	Texture	Moisture movement	Working qualities	Durability	Permeability	Uses
Redwood, European† *Pinus sylvestris* Scandinavia, Russia	Softwood Pale yellowish-brown to red-brown	510	Medium	Medium	Medium	Non-durable	Moderately resistant	Construction. Joinery. Furniture
Spruce, Canadian *Picea* spp. Canada	Softwood White to pale yellow	400/* 500	Medium	Small	Good	Non-durable	Resistant	Construction. Joinery
Spruce, Sitka† *Picea sitchensis* UK	Softwood Pinkish-brown	450	Coarse	Small	Good	Non-durable	Resistant	Construction. Packaging. Pallets
Spruce, Western white† *Picea glauca* N. America	Softwood White to pale yellow/ brown	400/* 500	Medium	Small	Good	Non-durable	Resistant	Construction. Joinery.
Wester red cedar† *Thuja plicata* N. Amercia	Softwood Reddish brown	390	Coarse	Small	Good	Durable	Resistant	Shingles, exterior cladding. Greenhouses. Beehives
Whitewood, European† *Picea abies* and *Abies alba* Europe, Scandinavia, Russia	Softwood White to pale yellowish-brown	470	Medium	Medium	Good	Non-durable	Resistant	Interior joinery. Construction. Flooring
Yew *Taxus baccata* Europe	Softwood Orange-brown to purple-brown	670	Medium	Small/ medium	Difficult	Durable	Resistant	Furniture. Turnery. Interior joinery

* Density can vary by 20% or more
† Structural properties included in BS 5268: Part 2: 1988

Index

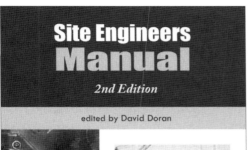